Issues in Higher Education

Series Editor: GUY NEAVE, International Association of Universities, Paris, France

Other titles in the series include

GOEDEGEBUURE et al.
Higher Education Policy: An International Comparative Perspective

NEAVE and VAN VUGHT
Government and Higher Education Relationships Across Three Continents: The Winds of Change

SALMI and VERSPOOR
Revitalizing Higher Education

YEE
East Asian Higher Education: Traditions and Transformations

DILL and SPORN
Emerging Patterns of Social Demand and University Reform: Through a Glass Darkly

MEEK et al.
The Mockers and Mocked? Comparative Perspectives on Differentiation, Convergence and Diversity in Higher Education

BENNICH-BJORKMAN
Organizing Innovative Research? The Inner Life of University Departments

HUISMAN et al.
Higher Education and the Nation State: The International Dimension of Higher Education

CLARK
Creating Entrepreneurial Universities: Organizational Pathways of Transformation.

GURI-ROSENBLIT
Distance and Campus Universities: Tensions and Interactions. A Comparative Study of Five Countries

TEICHLER and SADLAK
Higher Education Research: Its Relationship to Policy and Practice

TEASDALE and MA RHEA
Local Knowledge and Wisdom in Higher Education

TSCHANG and DELLA SENTA
Access to Knowledge: New Information Technology and the Emergence of the Virtual University

HIRSCH and WEBER
Challenges facing Higher Education at the Millennium

TOMUSK
Open World and Closed Societies: Essays on Higher Education Policies "in Transition"

SÖRLIN AND VESSURI
Knowledge Society vs. Knowledge Economy: Knowledge, Power, and Politics

SAGARIA
Women, Universities, and Change: Gender Equality in the European Union and the United States

SLANTCHEVA AND LEVY
Private Higher Education in Post-Communist Europe: In Search of Legitimacy

The IAU

The International Association of Universities (IAU), founded in 1950, is a worldwide organization with member institutions in over 120 countries. It cooperates with a vast network of international, regional and national bodies. Its permanent Secretariat, the International Universities Bureau, is located at UNESCO, Paris, and provides a wide variety of services to Member Institutions and to the international higher education community at large.

Activities and Services

- IAU-UNESCO Information Centre on Higher Education
- International Information Networks
- Meetings and seminars
- Research and studies
- Promotion of academic mobility and cooperation
- Credential evaluation
- Consultancy
- Exchange of publications and materials

Publications

- International Handbook of Universities
- World List of Universities
- Issues in Higher Education (monographs)
- Higher Education Policy (quarterly)
- IAU Bulletin (bimonthly)

Private Higher Education in Post-Communist Europe

In Search of Legitimacy

Edited by

Snejana Slantcheva

and

Daniel C. Levy

palgrave macmillan

International Association of Universities
International Universities Bureau

PRIVATE HIGHER EDUCATION IN POST-COMMUNIST EUROPE

Softcover reprint of the hardcover 1st edition 2007 978-1-4039-7425-9

First published in 2007 by
PALGRAVE MACMILLAN™
175 Fifth Avenue, New York, N.Y. 10010 and
Houndmills, Basingstoke, Hampshire, England RG21 6XS
Companies and representatives throughout the world.

PALGRAVE MACMILLAN is the global academic imprint of the Palgrave Macmillan division of St. Martin's Press, LLC and of Palgrave Macmillan Ltd. Macmillan® is a registered trademark in the United States, United Kingdom and other countries. Palgrave is a registered trademark in the European Union and other countries.

ISBN 978-1-349-53482-1 ISBN 978-0-230-60439-1 (eBook)
DOI 10.1057/9780230604391

Library of Congress Cataloging-in-Publication Data

Private higher education in post-communist Europe : in search of legitimacy / edited by Snejana Slantcheva and Daniel C. Levy.
p. cm.
Includes bibliographical references and index.

1. Education, Higher—Europe, Central. 2. Education, Higher—Europe, Eastern. 3. Private universities and colleges—Europe, Central. 4. Private universities and colleges—Europe, Eastern. I. Slantcheva, Snejana. II. Levy, Daniel C.

LA628.P68 2007
378.43—dc22 2006049477

A catalogue record for this book is available from the British Library.

Design by Newgen Imaging Systems (P) Ltd., Chennai, India.

First edition: February 2007

10 9 8 7 6 5 4 3 2 1

Transferred to digital printing in 2007.

Contents

Part II Country Perspectives

Part III Concluding Reflections

List of Tables

Acknowledgments

We owe a debt to the International Higher Education Support program of the Open Society Institute (HESP), Budapest, Hungary, especially to Rhett Bowlin, Director of HESP, and Voldemar Tomusk, Deputy Director. HESP's generous grant made possible a large international conference, which started us on the way toward this volume, as well as the subsequent preparation of the volume itself. The conference, June 2004 in Sofia, Bulgaria, was entitled "In Search of Legitimacy: Issues of Quality and Recognition in Central and Eastern European Private Higher Education." The chapters in the present volume are original texts, deriving from no prior versions other than the fresh conference papers.

We also acknowledge the vital role of the Program for Research on Private Higher Education (PROPHE), an acknowledgement that perforce means gratitude to the program's chief sponsor, the Ford Foundation, and the Head of the Foundation's Higher Education Research Unit, Jorge Balán. We further extend our gratitude to all contributors in this volume for their persistence and good spirit through rounds of exchanges with the editors. We are also grateful to Series Editor Guy Neave for his constant support and guidance. Last but not least, special thanks go to our families for their indulgent support.

List of Contributors

Pepka Boyadjieva is Professor at the Institute of Sociology at the Bulgarian Academy of Sciences. She is Vice-President of the Bulgarian Sociological Association. Boyadjieva has been lecturing at the University of Sofia and the New Bulgarian University and was awarded the Andrew Mellon Fellowship twice. Her research interests pertain to different models of relationship between university and society; education, science, and modernization; life long learning.

Hans C. Giesecke is President/CEO of the Independent Colleges of Indiana, Inc. Prior to this, he served as the advisor for student affairs at International University Bremen in Germany, as president of the Tennessee Independent Colleges and Universities Association (TICUA), and as director of marketing and research with the Association of Independent California Colleges and Universities (AICCU).

Julita Jablecka is Associate Professor at the Centre for Science Policy and Higher Education at Warsaw University, the author of about eighty publications in Polish and English. Her research interests include management of universities, research management and science policy.

Kevin Kinser, Assistant Professor in the Department of Educational Administration and Policy Studies, University at Albany, State University of New York. Kinser studies nontraditional and alternative higher education, particularly the organization and administration of for-profit and virtual universities, and the various ways in which institutions of higher education choose to serve and support students. Kinser is a core member of PROPHE.

Marek Kwiek is Director of the Center for Public Policy and Professor of Philosophy at Poznan University, Poland. His areas of interest include higher education policy, globalization and Europeanization, and transformations of the welfare state. He is the author or editor of eight books and eighty papers. Kwiek is a PROPHE Affiliate.

Daniel C. Levy is Distinguished Professor, SUNY. At the University at Albany-SUNY he holds appointments in multiple units, especially in educational policy. Levy directs PROPHE (Program for Research on Private Higher Education), a global scholarly network funded by the Ford Foundation. His nine books and over hundred articles concentrate on higher education policy, related nonprofit sectors, or Latin American politics. Levy has lectured at nearly all the top-ranked U.S. universities and in six continents, and has consulted widely.

Guy Neave is Professor of Comparative Higher Education Policy Studies at the Centre for Higher Education Policy Studies (CHEPS) Twente University, Netherlands and Director of Research at the International Association of Universities, Paris, France. A Foreign Association of the National Academy of Education of the United States of America, he has written far too much on higher education policy in Western Europe, quite apart from having been Joint Editor in Chief of the Pergamon Encyclopedia of Higher Education (1992), Editor of the Journal Higher Education Policy and series editor of Palgrave Macmillans series Issues in Higher Education. He lives in the Far West of the Paris Basin.

Luminiţa Nicolescu is Professor of International Marketing at Academy of Economic Studies in Bucharest, Romania. She has strong interests in the field of higher education as she extensively researched and published in the field, focusing mainly on Central and Eastern Europe and Romania.

Marie Pachuashvili is Ph.D. candidate in political science at Central European University in Budapest, Hungary and PROPHE Affiliate. Her current interests include post-communist regime transitions, higher education governance structure changes and private higher education development.

Oleksiy Panych is Vice-Director for Research at the Institute of Human Studies of Donetsk National University (Ukraine), and Adjunct Professor at the Department of Educational Leadership, Management and Policy of Seton Hall University (NJ, USA). His main fields of interest are history of Western philosophy and culture, private higher education and educational policy.

Robert D. Reisz is Associate Professor at the West University Timisoara in Romania and researcher at the Institute for Higher Education Research of the "Martin Luther" University Halle-Wittenberg. He has previously worked at the Hungarian Institute for Educational Research in Budapest, the Research Center for Higher Education and Work in Kassel, Germany and the New Europe College in Bucharest. His research has concentrated, on the one hand, on private higher education, and, on the other hand, on the statistical analysis of educational systems.

Peter Scott is Vice-Chancellor of Kingston University in London and was formerly Pro-Vice-Chancellor and Professor of Education at the University of Leeds. Until 1992 he was Editor of "The Times Higher Education Supplement." He is a member of the Boards of the European Universities Association (EUA) and the Higher Education Funding Council for England (HEFCE). His research interests include the governance and management of universities, the globalization of higher education and the development of new paradigms of research and knowledge production.

Snejana Slantcheva is Head of the Center for Research on Higher Education in Central and Eastern Europe, Blagoevgrad, Bulgaria, and Head of PROPHE's Regional Center for Central and Eastern Europe. Her research and publications focus on higher education policy issues in the countries of Central and Eastern Europe. She has edited a multi-authored volume and has published articles and book chapters on higher education policy, the quality of teaching and learning, the academic profession, private higher education, and the Bologna process.

Joseph Stetar is Professor in the Department of Educational Leadership, Management and Policy at Seton Hall University in South Orange, N.J., USA. His research interests include higher education reform in South Africa which has received support from The Andrew W. Mellon Foundation and the emergence of private higher education in the former Soviet Union which has received support from USAID, IREX and the US Department of State.

Dmitry Suspitsin, a core member of PROPHE, is completing his doctoral dissertation in Higher Education Administration and Comparative Education at The Pennsylvania State University. His dissertation research is supported by the International Dissertation Field Research Fellowship from the Social Science Research Council. Suspitsin's research interests include the worldwide emergence and development of private sectors in higher education and private universities' relationships with the nation-state, public universities, and other societal actors and institutions.

Andrew Tatusko is a candidate for the Ph.D. in Education Leadership, Management, and Policy at Seton Hall University, South Orange, N.J., USA, where his research interests are primarily in the relationship of sectarianism and religiously-affiliated higher education.

Introduction

Private Higher Education in Post-Communist Europe: In Search of Legitimacy

Snejana Slantcheva and Daniel C. Levy

Since 1990 we have witnessed unmatched growth of private higher education in most former communist countries.[1] Nowhere else has the change been as concentrated in time and as inclusive of so many countries that share a historical legacy.[2] Although private sector growth has been common worldwide, its development across Central and Eastern Europe is more striking in that it comes against the backdrop of at least four decades of communist public monopoly and historically limited higher education enrollment.

On the one hand, the growth of private higher education across Central and Eastern Europe is part of a larger, global process that involves an increase in the privatization or marketization of public services and a reduction of state control.[3] On the other hand, it is also part of the broader political-economic transformation of the post-communist countries. After the fall of the Berlin Wall, the political changes in Central and Eastern Europe signaled the beginning of large-scale economic transformation. A shift in patterns of ownership—from the state to private—was at the heart of this transformation. Indeed, "With the fall of the command economy in the former USSR, Eastern Europe and People's Republic of China, the post-communist societies became the testing ground for privatization on the massive scale" (Gupta 2000, p. 6).[4]

Finally, the emergence of private higher education in Central and Eastern Europe has gone hand in hand with a broader liberalization of higher education. One aspect of this liberalization is accommodation to a significant increase in demand for higher education overall. Another is the partial privatization of public universities and colleges with respect to alternative sources of funding, tuition fees, business ties, and support. Privatization within public institutions has increasingly blurred the boundaries between private and public institutions. While the private explosion epitomizes the "market" in higher education, the public sector is also much more market-oriented than before.

New institutions are generally met with mistrust by the public, and the case is no different for private higher education in the post-communist countries of Central and Eastern Europe and the former Soviet republics. Ever since their establishment, private colleges and universities in the region have had to struggle to gain social acceptance. However, the lack of robust legitimacy is not just the consequence of weak social standing of these institutions in the mind of a public unaccustomed to private higher education; it is also reflective of more fundamental problems of mass higher education in the region's transitional societies, including faltering educational standards, an influx of market-driven principles, the shifting character of national economies, and reduced financial support.

The chapters in the present volume address the growth and role of the private sectors of higher education in Central and Eastern Europe and former Soviet republics through the organizing theme of legitimacy. In presenting these chapters, we hope to accomplish four specific goals.

First, we hope to contribute to the still limited international literature on private higher education outside the United States. Our focus is on Central and Eastern Europe. Although statistical and descriptive information on the private sectors across this region has been increasing, this increase has been only gradual, and critical analysis has been rare. What is more, the private sectors across Central and Eastern Europe have been generally absent from European research programs and debates on the European Higher Education Area. Indeed, the Central and Eastern European region has not, to date, been adequately represented in the literature on private higher education worldwide. Thus, this volume endeavors to construct a picture of the region itself and to bring that picture into European and global considerations.

Second, the different chapters offer analytical perspectives on issues that are related to wider trends in contemporary higher education. Thus, this volume opens windows to the development of contemporary higher education systems across Central and Eastern Europe. Simultaneously, it connects private higher education to more general social, political, and economic trends.

Third, the volume aims to blend comparative and nation-specific perspectives. The first chapter (by Guy Neave) reviews the origins and development of the notion of "private" in higher education in the West of Europe. The next four chapters (by Snejana Slantcheva, Marie Pachuashvili, Hans C. Giesecke, and Marek Kwiek) consider the legitimacy of private higher education throughout Central and Eastern Europe. The ensuing six chapters take a nation-specific approach to the legitimacy of a given country's private sector; they deal with Romania, Russia, Poland, Bulgaria, and Ukraine. Kevin Kinser's discussion in chapter eleven focuses on the legitimacy of the U.S. for-profit sector as a sector that, to a large extent, faces similar challenges with private institutions across Central and Eastern Europe. The last two chapters, by Daniel C. Levy and Peter Scott, offer final reflections on issues related to private institutions of higher education across Central and Eastern Europe and beyond.

Fourth, the volume should contribute to the broader social, economic, and political literature on post-communism and Central and Eastern Europe. As Holmes (1997) notes, there is a great need to assess post-communist reality and its regional setting. This volume's contribution comes through intensive and cross-national analysis of one sector, higher education, notably private higher education. Of course, this is not to minimize the importance of more general country reviews, but a complementary single-sector approach can help build from the bottom up. Authors of the country-specific chapters are scholars within their countries with knowledge of general tendencies.

Together, then, the chapter authors seek to present an informed picture of the private higher education sector phenomenon across Central and Eastern Europe. They contribute data, define concepts, clarify distinctions, and explore different aspects of organizational legitimacy.

This introductory chapter proceeds to provide context for the chapters that follow. It sketches the contours of the book's dual focus: the private higher education sectors of Central and Eastern Europe and their legitimacy.

Growth and Development

While the historical forces inducing change after 1989 were comparatively homogenous across Central and Eastern Europe, significant variations exist with respect to private higher education when we consider individual countries. The growth of private sectors across the post-communist countries has been powerful, but not even. Different national patterns of growth have been influenced by historical enrollment rates, speed of reforms, social values, entrepreneurship, and the strength of civil society. At one extreme, Turkmenistan is a rare case where private institutions have yet to appear (Tursunkulova 2005). In Croatia and the Slovak Republic, private institutions educate as few as 3.0 and 4.6 percent (for academic 2004/2005) of the countries' student populations, respectively. At the other extreme, private sectors in Estonia, Poland, and Romania enroll almost one-third of all students. Other countries such as Bulgaria, Hungary, and Russia have more moderate private student enrollments, around 15 percent (table I.1 in Introduction Appendix).

Private higher education has had little history or resonance in modern Europe. In many Western European countries, private sectors have existed for a long time but play a marginal role. The United Kingdom, France, Denmark, Ireland, and the Scandinavian countries are examples of countries with low student percentage enrollments, ranging between under 1 and 15 percent, often in what Geiger (1986) calls "peripheral" sectors, including religious institutions. In Greece, private higher education institutions are still not formally recognized. Recently, however, notable private creation or growth has occurred in Austria, Germany, Italy, Portugal, and Spain, as well as in Turkey and Israel.

The comparatively more powerful creation and growth of private sectors across Central and Eastern Europe since 1989 have often followed a common pattern of development over time: the initial explosive growth has been followed by a decade of relative stability. In Poland and Romania, for example, the initial explosion of private institutions approached 25 percent of student enrollments within the first five years, and in Bulgaria, Hungary, and Russia—12 percent. Yet these private enrollment percentages have not changed much since the mid-1990s. This relative proportional stability should not imply a stagnation of private higher education across Central and Eastern Europe since higher education systems in general grew substantially in the 1990s. This stability instead reveals strong increases in private enrollment.

Furthermore, private shares have remained steady in the face of recent enrollment stagnation due, above all, to negative demographic tendencies. Enrollment stagnation has not been limited to Central and Eastern Europe. In many Western European countries, the increase in the number of students has been contained recently, while student enrollment levels in France, Italy, and Austria have remained unchanged or decreased (Eurydice 2002).

Legitimacy of Higher Education Institutions

The notion of legitimacy refers broadly to the legal and social acceptability of an institution in society. Legitimacy is not secured simply and solely through the authority of law; it is also rooted in social norms and values. "Legitimacy is a conferred status and, therefore, always controlled by those outside the organization" (Pfeffer and Salancik 1978, p. 194). In complex environments, organizations have multiple sources of legitimacy. Many actors may participate in the legitimation process. Those with authority to confer legitimacy formally or informally evaluate aspects of an organization.

They exert varying degrees of influence on the overall levels of legitimacy. This is no less true for institutions of higher education.

Burton Clark's identification of the state, the market, and academia as the triangle of power (1983) can be utilized as highlighting three major sources of legitimacy in higher education. Shifts are apparent in all three and in their interactions. The new state is of course fundamentally different from the state under communism, far from being the sole legitimating authority. Still, the state is of core significance in higher education across Central and Eastern Europe, even for private higher education. In most countries, the appearance of private institutions of higher education took the state and society by surprise. This often meant private proliferation amid little state regulation. "Delayed state regulation" over the private sector, common in much of the world (Levy 2006b), sharply characterized the region. Increased regulation has since been quite variable by country. State regulation covers aspects ranging from authorization of new institutions to daily functioning. The state can also affect institutional standing and legitimacy through multiple related actors (e.g., accreditation bodies). As shown in the Ukrainian chapter, for example, the state can also both strengthen and call to question the legitimacy of institutions.

The market may also be viewed as an important source of legitimacy. Several of our authors note that the market is taken as an alternative source of legitimacy for private institutions strongly geared to it. Indeed it appears that the market is an increasingly important source of legitimacy for even the (evolving) public sector. However, the concept of legitimacy is trickier when applied to the market as opposed to the state. The market is more about success on "instrumental" or "pragmatic" terms rather than on normative legitimacy terms as typically specified and understood.

The legitimizing role of academia has also been changing and controversial along with that of the state and market. In many instances, the state and public universities themselves have been quite active in the creation of private higher education institutions, as Suspitsin shows for Russia. In most cases, however, strong resistance can be documented through many Central and Eastern European countries—as Jablecka illustrates from Poland—where academics from public universities may work through the state to impose rules and categories on private institutions in ways that deny them certain kinds of legitimacy. Furthermore, public institutions are not the only ones trying to question the legitimacy of private higher education institutions. Leading private universities have reasons to join in, sometimes even try to block aspiring new private entrants, at least for accreditation. Reasons include genuine conviction about low quality, fear of competition and anxiety that the addition of low legitimacy institutions may reflect to their own detriment as views of the private sector become more negative. Sometimes nonprofit institutions deny legitimacy to for-profit institutions based on the idea that for-profit institutions serve solely the market while nonprofit institutions are more like the public institutions in social mission.

Different organizations may draw from all, most, some, or just a few sources of legitimacy. This is sometimes a matter of choice, sometimes of necessity, and often a mix of the two. In any event, the *number* of sources is not fully correlated with the *amount* of legitimacy received from sources; in fact, Meyer and Scott (1983, p. 202) argue that the "legitimacy of a given organization is negatively affected by the number of different authorities sovereign over it and by the diversity or inconsistency of their accounts of how it is to function." Some sources provide much more legitimacy than others. This depends not only on the source but also on the recipient. Thus, a source dynamic that provides ample legitimacy to one institution may have little impact on another. Much here turns on the nature of the institution or sector, as well as on broader matters of the context within which higher education functions. At the same time, not all institutions require the same aggregate contribution from sources to have adequate legitimacy. As noted by several of our authors (for example, Reisz), market ease or pragmatic prowess may be vital to survival and possibly great success even when conventionally defined legitimacy from external actors is limited. Clearly, private and public sectors differ in their sources of legitimacy. Even when they count on common sources, private higher education institutions rely to different degrees on specific sources.

Challenges to Legitimacy

A focus on legitimacy is logical and important, given the time and place covered in this book. Revolution shattered key prior bases of legitimacy and requires new ones, a major point reviewed in Scott's chapter. Whether and how legitimacy could be transformed for the new day is at the heart of post-communist transformation. The question is as pertinent to higher education as to other fields.

New institutions often face challenges to legitimacy. They lack tradition, social standing, established support, and secure sustenance. Their norms may be not only new but even seen as contradictory to socially ingrained ones. As Lipset once remarked, "a crisis of legitimacy is a crisis of change" (1981, p. 65).

Several factors have come together to impose strong challenges for private higher education in post-communist countries, even though the sector avoids other delegitimizing factors associated with the public higher education's communist legacy. As new institutions, which in many countries instantly mushroomed in a few years, private universities and colleges drew special public scrutiny. Would societies accustomed to only public institutions find private ones legitimate? Could such legitimacy come through emulating the known public forms? If emulation works, could private higher education gather the resources, will, and other necessities to copy public counterparts and to compete effectively with them? On the other hand, where emulation is neither feasible nor desired, the natural vulnerability of new forms is stark. There is no assurance that private higher education would have the support to go that route, or to stick with it amid difficulties.

The lack of long-standing traditions of private higher education across the countries of Central and Eastern Europe combines with the general public mistrust of market forces in education. The norm that the state is responsible for the provision of higher education has been widely held both before and after the period of communism. A related factor negatively affecting social acceptance has been the lack of tradition or familiarity with the idea of nonprofit institutions (and most private higher education institutions across Central and Eastern Europe are nonprofit).[5] At the same time, for-profit is even a less legitimized concept than for example in North America, making it difficult to imagine the sort of vibrant for-profit sector that Kinser's chapter reports for the United States.[6]

In the early post-communist years, these challenges were partly offset by several striking opportunities. Something new is not automatically rejected when the old is so discredited (Levy 2006a). Or the old may simply be inadequate to handle a new and expanded situation by itself. Both points have been pertinent to the post-communist rise of a private higher education sector. Thus, an atmosphere of hope could be friendly to new structures and initiatives and the soaring demand for higher education presented a market for providers. Even a modicum of legitimacy could be enough if students had no institutional alternative.

Over time, however, conditions changed and the legitimacy of private institutions was increasingly questioned. A major weakness has been the strong perception that private institutions are not academically serious. Tomusk (2003) highlights how an early post-communist reformist thrust seeking a viable new academic model has wilted. Linked to that are perceptions that even many formally nonprofit institutions basically pursue financial gain, as Nicolescu's chapter shows for Romania. The legitimacy deficit of private institutions relates to each of these challenges. Public trust has been problematic. It is not that private institutions were accepted in 1990 and not a decade later; in fact in some important ways legitimacy has increased, as Levy's chapter shows. It is that the challenges to legitimacy have evolved and the passage of time has not shown a unilinear trend to increased or secure legitimacy.

Whereas such challenges might be weathered, were the higher education enrollment boom to continue, the stagnation in student enrollment across the region threatens to outweigh the other challenges and exacerbate them. It also threatens to trump the opportunities open to the private sector. Furthermore, the partial privatization of public universities represents a major threat to the private sector's claims of pragmatic, market legitimacy. A leading example is the acceptance by public universities of students who could not be admitted under the quota for subsidized students; these additional students are fee-paying. Thus, intersectoral competition has intensified.

Some of the challenges identified above help explain why the growth of private sectors has not been greater and, in particular, why certain types of higher education do not flourish in the private sector. They can also help explain the decline and even demise of particular private higher education institutions, as exemplified in countries such as Romania.

Opportunities for Addressing Legitimacy Deficits

Private institutions often have some room within which to strive for legitimacy. A good example can be seen with regard to state regulation, of which accreditation is a particularly important manifestation. Accreditation and other state regulations are of course major challenges to many private higher education institutions, sometimes insurmountable. Private institutions often complain about unfair accreditation criteria and enforcement, and the complaints are largely about the poor fit with the institutions' missions and strategies for their own sort of legitimacy. Yet the same regulations may play a central role for the broad legitimation of private institutions. Private institutions that receive accreditation acquire an official stamp of legitimacy. So those institutions that are capable of accreditation may eagerly pursue it, with the added benefit of distinguishing themselves from the incapable institutions and thus gaining market share. For the leaders, a "virtuous circle" emerges, as performance leads to some legitimacy, which then is a tool to improve or expand performance, perhaps acquiring consequential opportunities for rights, such as offering advanced degrees, and benefiting as a result from increased societal interest and contact. Even short of accreditation, initial licensing and recognized compliance to state regulations yield some legitimation. Finally, though in still rare instances, state funding for private institutions, both of religious and nonreligious type, may be made available, reflecting and enhancing broad public legitimacy.

In their search for legitimacy, probably the most common and weighty claim of private institutions concerns access, as Pachuashvili's chapter notes. Prior to communism, most Central and Eastern European countries had elite higher education systems. The political changes of 1989 unleashed popular demand for higher education. The state was either unable or unwilling to finance the totality of expansion and it was unable or unwilling to thwart the demand. In this context, newly founded private institutions complemented the expansion of public higher education, while the state in effect surrendered its traditional legitimacy claim of being the sole provider of higher education opportunity. Furthermore, the private sector's access claim has gone hand in hand with an equity claim, as expansion opens opportunities for less privileged groups (especially since public universities remain the first choice of the most privileged students, as many of our authors show). Considering how much of a challenge to private institutions' legitimacy clusters around the charges that they serve private over public interests and privilege over inclusiveness, the access/equity claim is crucial.

Yet legitimacy is pursued not only through concentration on something particular or through distinctiveness. Copying is also a very common route. Emulating legitimate established practices and forms might bolster institutional legitimacy, especially when the system is expanding. Neo-institutionalists emphasize the importance of *social fitness* or, in other words, the acquisition of a form regarded as legitimate in a given institutional environment. DiMaggio and Powell (1983) call attention to coercive, normative, and mimetic "isomorphism" that make organizations more similar without necessarily making them more efficient. Thus, the isomorphic tendencies of Central and Eastern European private higher education institutions result in part from a lack of imagination or care whereas copying can be a conscious strategy as well. Regionally powerful examples of isomorphism attributable to both a lack of alternatives or innovations and strategic purpose include the private sector's widespread employment of public university professors and use of public curricula.

Different mixes of emulation and innovation, common broad approaches and niche searching, placid policy and strategy are found in different private institutions struggling to overcome legitimacy deficits and challenges. Legitimacy is pursued in varying ways with varying degrees of success. At the extremes, substantial legitimacy has been earned by a few academically strong and/or large universities, while the majority of institutions survive in a more precarious state, fulfilling an access and perhaps labor market function, sometimes without formal licensing or recognition.

Overview of the Book Chapters

The chapters in the book address a host of these issues and are organized in three parts. Part One—Regional Perspectives—includes chapters that consider the legitimacy of private higher education from a regional perspective. In chapter one, Guy Neave examines the drive toward the privatizing of higher education within Western Europe. He maintains that both the pattern of development and the issues privatization poses in this setting appear as a species of exception, when viewed from outside Europe and within the imperatives of privatization itself. Europe was the birthplace of the university and of the Nation State. Today, the path to privatization in higher education is pursued with both caution and reticence. Furthermore, Europe's move toward both a Higher Education Area and a European Research Area follows a rationale which, if not incompatible with privatization, is very different from it.

In chapter two, Snejana Slantcheva explores the reasons behind the legitimacy deficits of private higher education institutions in Central and Eastern Europe. She argues that, to a large extent, the search for institutional legitimacy reflects the transition in existing values in post-communist societies, which are in general suspicious of private provision of higher education, question the prioritization of human resource development over scientific research, and have historically placed more reliance on the state to be the caretaker of private goods.

In chapter three, Marie Pachuashvili finds that differences in the legitimacy sources for private higher education have produced diverse growth and development patterns as well as multiple types of organizational legitimacy in countries across Central and Eastern Europe and former Soviet republics. Notwithstanding the importance of organizational legitimacy in sectoral growth in general, the post-communist evidence demonstrates that private institutions can prosper even when largely lacking formal legitimacy. Easy expansion can even thwart the building of institutional legitimacy. Subsequent strict regulatory measures adopted by some countries impede organizational growth but also confer legitimacy.

In chapter four, Hans C. Giesecke examines the factors that affect the perceived movement of newly founded institutions of private higher education in Central and Eastern Europe along an identifiable legitimacy continuum encompassing processes for attaining effectiveness and viability. Through a web-based perceptional survey amongst institutional rectors in Hungary and Poland and by distilling the observations of hundreds of students from Central and Eastern Europe, Giesecke formulated a rank-ordered listing of the most important legitimacy factors. The International

University Bremen in Germany is used as a comparative case to illustrate the path of one new private institution in Europe along the legitimacy continuum.

In chapter five, Marek Kwiek argues that the Bologna process seems to disregard the rise of the private sector in higher education across Central and Eastern Europe. As a result, the ideas behind the Bologna process and the analytical tools it provides may have unanticipated effects on higher education systems there. At the same time, the expansion of educational systems throughout Central and Eastern Europe is crucial for the implementation of the Lisbon strategy of the EU and, more generally, the creation of the "Europe of Knowledge."

Part Two—Country Perspectives—of the book includes chapters that look more closely at individual national private higher education sectors and discuss issues related to their legitimacy. The countries included are: Romania and Poland as examples of countries with large private higher education sectors; Bulgaria, Russia, and Ukraine representing countries with mid-size private higher education sectors. In chapter six, Robert D. Reisz analyzes the legitimating discourse of private higher education institutions in Romania as expressed in their mission statements. The author offers a categorization of private institutions according to their legitimacy discourse and argues that Romanian private institutions, like private institutions in other countries of the world are challenged by the duality of the liability of newness and the liability of privateness in their search for legitimacy.

Using institutional founders as indicators of the sources of sponsorship and legitimacy available to nonstate institutions, in chapter seven Dmitry Suspitsin lays out several distinct legitimacy-building orientations of nonstate universities in Russia and offers partial explanations for the success of these strategies. Ranging from governmental organizations to state universities to private actors, the founders in effect represent continua of privateness and point to varying degrees of proximity to either state-run or private organizations on the part of nonstate institutions.

In chapter eight, Julita Jablecka focuses on the influence and attitudes of the public university academic community toward the nonpublic institutions in Poland. The chapter also traces the gradual legal equalization of the public and nonpublic sectors and the growing legal legitimacy of the nonpublic institutions.

Luminita Nicolescu, in chapter nine, analyzes the legitimacy of private higher education in Romania and argues that the overall legitimacy-gaining strategy of the private higher education sector in Romania has been based on copying the public sector in order to minimize risks. The high degree of isomorphism between public and private in academic respects has been further pushed by the commonly set standards for the whole Romanian higher education system, notably standards based on the norms met in the public sector. At the same time large differences between the two sectors in terms of economic goals are still present.

In chapter ten, Pepka Boyadjieva and Snejana Slantcheva argue that public perceptions of Bulgarian private universities and colleges are constructed against the traditional image of academic organizations. The authors focus on public perceptions of the private sector of higher education as a barometer of the social legitimacy of Bulgarian private institutions of higher education.

In chapter eleven, Joseph Stetar, Oleksiy Panych, and Andrew Tatusko argue that Ukrainian private higher education institutions are caught within a paradoxical tension between legitimating and delegitimating trends and structures. This tension is particularly evident in state policies of accreditation, licensing, and taxation. Avenues to resolve the paradox are through finance, social capital, broader cooperation between state and private institutions where private institutions can build upon the established legitimacy of the state institutions, and through affiliations and stratification.

In chapter twelve, Kevin Kinser looks at a multidimensional model of legitimacy proposed by Suchman (1995) and applies it to the private, for-profit sector in the United States. Conflicting legitimacy assessments of the sector by various stakeholders are highlighted in a recent (and as yet unresolved) policy debate in the United States that centers on the role of for-profit higher education in the reauthorization of the Higher Education Act. For-profit legitimacy is proposed as a threshold question: do for-profit colleges and universities have enough legitimacy to support an affirmative policy environment?

Part Three—Concluding Reflections—of the book comprises two chapters that offer final reflections on the issues explored in this volume. In chapter thirteen, Daniel C. Levy reviews and brings together major factors that have threatened and limited private higher education legitimacy. He then identifies major factors that have brought new opportunities, with new sources and types of legitimacy. Tied to the broader political, economic, and social context, the more and less favorable factors lead Levy to the conclusion that legitimacy has fresh bases but remains circumscribed and precarious.

In chapter fourteen, Peter Scott explores the question whether the legitimacy challenges before post-communist private institutions of higher education can be attributed to specific historical circumstances or are better explained in terms of the wider crisis of legitimacy experienced by most post-public higher education systems. The chapter analyzes the impact of the transition from communist to post-communist societies on policy experiments and the private sector, and the origins, impact and significance of private institutions within the evolution of post-communist higher education.

Appendix

Table I.1 Student enrollments in countries of Central and Eastern Europe and former Soviet republics

Country	*Student enrollments*							
	Academic 2003–2004				*Academic 2002–2003*			
	Public	*%*	*Private*	*%*	*Public*	*%*	*Private*	*%*
Armenia	55900	70.3	22000	29.7	54100	74.8	18200	25.2
data 2004–5	62500	73.4	22600	26.6				
Azerbaijan	104000	85.6	17500	14.4	101700	84.75	18300	15.25
*Belarus	279300	82.66	58600	17.34	272900	85.10	47800	14.90
data 2004–5	304300	83.81	58800	16.19				
data 2005–6	325100	84.79	58300	15.21				
Bulgaria	195666	85.64	32802	14.36	199529	86.56	30984	13.44
data 2004–5	198810	83.57	39099	16.43				
*Czech Republic	281312	92.3	23561	7.7	259334	93.8	17006	6.2
data 2004–5	298754	91.1	29201	8.9				
Estonia	52331	79.70	13328	20.30	50709	79.70	12916	20.30
data 2004–5	53390	78.79	14370	21.21				
*Georgia	123900	80.82	29400	19.18	122200	79.50	31500	20.50
*Hungary	351154	85.84	57921	14.16	327456	85.82	54101	14.18
data 2004–5	363961	86.34	57559	13.66				
Kazakhstan	358700	54.51	299400	45.49	338800	56.7	258700	43.3
data 2004–5	400000	53.54	347100	46.46				
Kyrgyz Rep.	187900	92.56	15100	7.44	184900	92.87	14200	7.13
data 2004–5	202500	92.76	15800	7.24				
Latvia	94368	74.45	32388	25.55	91646	77.11	27199	22.89
data 2004–5	94212	72.09	36481	27.91				
Lithuania	158799	93.0	11918	7.0	139244	95.5	6540	4.5
data 2004–5	176322	92.5	14379	7.5				
The FYR of Macedonia	44331	91.92	3896	8.08	46637	97.08	1402	2.92
data 2004–5	43293	91.68	3928	8.32				
Poland	1306225	70.5	545926	29.5				
data 2004–5	1337051	69.7	580242	30.3	1271728	70,6	528820	29.4
*Romania	476881	76.82	143904	23.18	457259	76.68	139038	23.32
Russia	5596000	86.67	860000	13.33	5229000	87.91	719000	12.09
data 2004–5	5860000	85.12	1024000	14.88				
Slovak Rep.	156651	97.83	3479	2.17	152705	99.12	1348	0.88
data 2004–5	169506	95.38	8208	4.62				
Ukraine	n/a		n/a			89.52	238,1	10.48

Student enrollments											
Academic 2001–2002				*Academic 2000–2001*				*Academic 1999–2000*			
Public	%	*Private*	%	*Public*	%	*Private*	%	*Public*	%	*Private*	%
47400	72.3	18200	27.7	43600	71.8	17100	28.2	39770	64.4	21992	35.6
99000	82.16	21500	17.84	91000	76.02	28700	23.98	88500	76.23	27600	23.77
260000	86.16	41760	1384	2,45100	87.00	36630	13.00	228600	87.22	33500	12.78
199716	87.44	28678	12.56	2,19067	88.69	27939	11.31	233907	89.51	27414	10.49
242922	95.1	12393	4.9	230347	95.3	11465	4.7	209708	95.1	10857	4.9
47374	78.42	13035	21.58	43171	76.49	13266	23.51	37173	74.98	12401	25.02
115500	78.35	31900	21.65	105800	76.16	33100	23.84	95000	70.31	40100	29.69
300360	85.99	48941	14.01	283970	86.76	43319	13.24	266144	87.06	39558	12.94
330800	64.27	183900	35.73	313800	71.2	126900	28.8				
191900	92.53	15500	7.47	174500	92.43	14300	7.57	146000	91.71	13200	8.29
88239	79.85	22261	20.15	86671	85.58	14599	14.42	77620	86.72	11890	13.28
114429	97.6	2861	2.4	97843	98.7	1297	1.3	84282	99.9	63	0.1
45624	98.07	918	1.97	44710	97.98	923	2.02	40246	97.74	932	2.26
1211379	70.4	509279	29.6	1119201	70.3	472340	29.7				
435406	74.78	146815	25.22	382478	71.74	150674	28.26	322129	71.17	130492	28.83
4797000	88.39	630000	11.61	4271000	90.08	471,000	9.92	3,728,000	91.53	345,000	8.47
145972	99.60	581	0.40	133993	99.22	1051	0.78	131715	99.37	829	0.63
1911,3	90.61	197,9	9.39	1770,9	91.71	160,1	8.29	1649,8	92.22	139,2	7.78

Table I.2 Number of institutions of higher education in countries of Central and Eastern Europe and former Soviet republics

Country	*Number of Institutions of Higher education*							
	Academic 2003–2004				*Academic 2002–2003*			
	Public	*%*	*Private*	*%*	*Public*	*%*	*Private*	*%*
Armenia	20	21.5	73	78.5	20	21.7	72	78.3
data 2004–5	20	22.7	68	77.3				
Azerbaijan	27	64.29	15	35.71	26	63.41	15	36.59
*Belarus	43	72.88	16	27.12	44	75.86	14	24.14
data 2004–5	43	78.18	12	21.82				
data 2005–6	43	78.18	12	21.82				
Bulgaria	37	72.55	14	27.45	37	72.55	14	27.45
data 2004–5	37	69.81	16	30.19				
Croatia								
data 2004–5	91	85.85	15	14.15				
*Czech Rep.	28+113	61.8	28+59	38.2	28+112	61.9	27+59	38.1
data 2004–5	26+116	59.9	36+59	40.1				
Estonia	22	46.80	25	53.20	22	44.90	27	55.10
data 2004–5	22	46.83	24	53.17				
*Georgia	26	14.78	150	85.22	26	14.45	154	85.55
*Hungary	31	45.59	37	54.41	30	45.45	36	54.55
data 2004–5	31	44.93	38	55.07				
Kazakhstan	46	25.56	134	74.44	50	28.25	127	71.75
data 2004–5	51	28.18	130	71.82				
Kyrgyz Rep.	31	65.96	16	34.04	31	67.39	15	32.61
data 2004–5	33	67.35	16	32.65				
Latvia	29	60.42	19	39.58	28	62.22	17	37.78
data 2004–5	36	64.29	20	35.71				
Lithuania	31	64.6	17	35.4	30	69.8	13	30.2
data 2004–5	31	64.6	17	35.4				
Macedonia	3	37.50	5	62.50	2	50	2	50
data 2004–5	3	37.50	5	62.50				
Poland	126	31.5	274	68.5	125	33.2	252	66.8
data 2004–5	126	29.5	301	70.5				
*Romania	55	45.08	67	54.92	55	44.00	70	56.00
Russia	654	62.52	392	37.48	655	63.04	384	36.96
data 2004–5	662	61.81	409	38.19				
Slovak Rep.	23	85.19	4	14.81	23	92.00	2	8.00
data 2004–5	23	82.14	5	17.86				
Ukraine	821	81.36	188	18.64	822	82.44	175	17.56
data 2004–5	764	79.09	202	20.91				

Number of Institutions of Higher education											
Academic 2001–2002				*Academic 2000–2001*				*Academic 1999–2000*			
Public	%	*Private*	%	*Public*	%	*Private*	%	*Public*	%	*Private*	%
20	22	71	78	19	21.1	71	78.9	16	16.3	82	83.7
25	62.5	15	37.5	25	58.14	18	41.86	25	59.52	17	40.48
44	75.86	14	24.14	43	75.44	14	24.56	42	73.68	15	26.32
39	78	11	22	39	86.67	6	13.33	39	86.67	6	13.33
+109	64.6	58	35.4	28+109	67.8	57	32.2	28+109	70.3	NA+58	29.7
23	46.94	26	53.06	23	50	23	50.00	21	51.22	20	48.78
26	14.53	153	85.47	26	15.12	146	84.88	24	12.9	162	87.10
30	46.15	35	53.85	30	48.39	32	51.61	30	33.17	59	66.29
59	31.89	126	68.11	58	34.12	112	65.88				
32	66.67	16	33.33	30	66.67	15	33.33	26	66.67	13	33.33
20	58.82	14	41.18	20	60.61	13	39.39	19	57.58	14	42.42
22	62.9	13	37.1	19	73.1	7	26.9	15	93.8	1	6.2
2	66.67	1	33.33	2	66.67	1	33.33	2	66.67	1	33.33
124	34	241	66.0	115	37.1	195	62.9				
57	45.24	69	54.76	59	46.83	67	53.17	58	47.93	63	52.07
621	61.6	387	38.4	607	62.9	358	37.1	590	62.83	349	37.17
23	95.83	1	4.17	21	91.30	2	8.70	21	91.30	2	8.70
812	82.6	171	17.4	816	83.35	163	16.65	809	83.32	162	16.68

Table I.3 Population and population density in countries across Central and Eastern Europe and former Soviet republics

Country	*Population*	*Area (km^2)*	*Density (Pop per km^2)*
Albania	3,129,678	28,748	109
Armenia	3,016,312	29,800	101
Azerbaijan	8,410,801	86,600	97
Belarus	9,755,106	207,600	47
Bosnia and Herzegovina	3,907,074	51,197	76
Bulgaria	7,725,965	110,912	70
Croatia	4,551,338	56,538	81
Czech Republic	10,219,600	78,866	130
Estonia	1,329,697	45,100	29
Georgia	4,474,404	69,700	64
Hungary	10,097,730	93,032	109
Kazakhstan	14,825,110	2,724,900	5.4
Kyrgyzstan	5,263,794	199,900	26
Latvia	2,306,988	64,600	36
Lithuania	3,431,033	65,300	53
Moldova	4,205,747	33,851	124
Montenegro	630,548	14,026	45 (figures from CIA World Factbook;Montenegro as of 2004)
Poland	38,529,560	312,685	123
Republic of Macedonia	2,034,060	25,713	79
Romania	21,711,470	238,391	91
Russia	143,201,600	17,098,242	8.4
Serbia	9,396,411	88,361	106 (figures from CIA World Factbook; Serbia as of 2002)
Slovakia	5,400,908	49,033	110
Slovenia	1,966,814	20,256	97
Tajikistan	6,506,980	143,100	45
Turkmenistan	4,833,266	488,100	9.9
Ukraine	46,480,700	603,700	77
Uzbekistan	26,593,120	447,400	59

Source: United Nations World Population Prospects (2004 revision). Data is for 2005, at http://en.wikipedia.org/wiki/List_of_countries_by_population_density, accessed October 3, 2006.

Data Sources

Armenia: National Statistical Service of the Republic of Armenia at http://www.armstat.am/ and its annual statistical reports at:
http://www.armstat.am/StatData/taregirq_05/taregirq_05_7.pdf
http://www.armstat.am/Publications/2004/soc_book/soc_book_3.pdf
http://www.armstat.am/Publications/2001/Armenia2001-eng/Armenia-3.pdf
http://www.armstat.am/Publications/2003/Armenia-2002/Armenia-02-III.3.1.3.pdf

Azerbaijan: The State Statistical Committee of the Azerbaijan Republic (at http://www.azstat.org/), Statistical Yearbook of Azerbaijan 2004 at http://www.azstat.org/publications/yearbook/SYA2004/Pdf/08en.pdf

Belarus: Галоўны інфармацыйна-аналітычны центр Міністэрства адукацыі Рэспублікі Беларусь. Статыстычны даведнік. Вышэйшыя навучальныя установы Рэспублікі Беларусь па стану на пачатак 2005/06 навучальны год. Мінск, 2005. (Главный информационно-аналитический центр Министерства образования Республики Беларусь. Статистический спавочник. Высшие учебные заведения Республики Беларусь. Минск, 2005) (General Information Analytical Centre for the Ministry of Education Republic of Belarus. The statistical reference book "Higher Educational Institutions in Belarus". Minsk, 2005).

*Note 1: Public and state institutions of higher education are combined.

*Note 2: Enrolments at public and state institutions of higher education are combined.

Bulgaria: National Statistical Institute, Education in the Republic of Bulgaria, Sofia: National Statistical Institute, 2004, 2005.

Czech Republic: Czech Statistical Office. 2006. Statistical Yearbook of the Czech Republic 2005 (in the Czech language). Retrieved March 23, 2006, from http://www.czso.cz/csu/edicniplan.nsf/p/10n1-05.

Czech Statistical Office. (2005). Statistical Yearbook of the Czech Republic 2004 (in the Czech language). Retrieved March 23, 2006, from http://www.czso.cz/csu/edicniplan.nsf/p/10n1-04.

Czech Statistical Office. (2004). Statistical Yearbook of the Czech Republic 2003 (in the Czech language). Retrieved March 23, 2006, from http://www.czso.cz/csu/edicniplan.nsf/p/10n1-03.

*Note 1: Table I.1 combines students enrolled in universities and tertiary professional schools.

*Note 2: In Table I.2, public institutions include public + state + regional; private institutions include religious professional tertiary schools. The first private universities were established in the summer of 1999 but they appear in statistical data for the first time in academic 2000/2001.

Croatia: State Bureau of Statistics, OECD Thematic Review of Tertiary Education (to be published in 2006).

Estonia: Source: Statistical Office of Estonia, Statistical Database Available at http://www.stat.ee Note: included are also private vocational education institutions that offer higher education.

Georgia: Ministry of Economic Development of Georgia, State Department for Statistics. Statistical Abstract, Tbilisi, 2003, 2004.

*Note: The Figures for the number of students approximate the hundreds.

Hungary: Statisztikai Tájékoztató. Felsőoktatás. Budapest. 2005. Oktatási Minisztérium (Statistical Information. Higher Education. Budapest. 2005. Ministry of Education).

*Note 1: The number of private institutions includes also church-run institutions.

*Note 2: Student numbers include full-time and part time students.

Kazakhstan: Agency of Statistics of the Republic of Kazakhstan at http://www.stat.kz/stat/index.aspx?sl=news&l=en; its report on "Современная система образования и профессиональной подготовки населения" at http://www.stat.kz/stat/index.aspx?p=analit-agent-2005&l=ru.

Kyrgyz Republic: National Statistics Committer of Kyrgyz Republic, Kyrgyzstan in Numbers: Statistical Documents, Bishkek, 2001, 2005.

Latvia: Pārskats par Latvijas augstskolu darbību 1999.gadā; Izglītības un zinātnes ministrija Augstākās izglītības un zinātnes departaments; 1999.

Pārskats par Latvijas augstskolu darbību 2000.gadā; Izglītības un zinātnes ministrija Augstākās izglītības un zinātnes departaments; 2000.

Pārskats par Latvijas augstskolu darbību 2001.gadā; Izglītības un zinātnes ministrija Augstākās izglītības un zinātnes departaments; 2001.

Pārskats par Latvijas augstskolu darbību 2002.gadā; Izglītības un zinātnes ministrija Augstākās izglītības un zinātnes departaments; 2002.

Pārskats par Latvijas augstāko izglītību 2003.gadā (skaitli, fakti, tendences); Izglītības un zinātnes ministrija Augstākās izglītības un zinātnes departaments; 2003.

Pārskats par Latvijas augstāko izglītību 2004.gadā (skaitli, fakti, tendences); Izglītības un zinātnes ministrija Augstākās izglītības un zinātnes departaments; 2005.

Lithuania: Statistics Lithuania.

The FYR of Macedonia: The Ministry of Education and Science—the Board of Accreditation. Note: Data also includes part-time students.

Poland: GUS, Szkoщy wyэsze I ich finance w roku 2000/1, 2001/2, 2002/3, 2003/4, 2004/5 (Higher schools and their finances, different years) Warsaw.

Romania: Institutul National de Statistica (National Statistical Institute), Anuarul Statistic al Romaniei, 2004 (Statistical Yearbook of Romania, 2004), 2005, at http://www.insse.ro/anuar_2004/asr2004.htm

*Note: Data from the National Statistical Institute of Romania include all private higher education institution, including not accredited and not authorized ones and differ as such from data of the Ministry of Education or the National Committee for Academic Evaluation and Accreditation.

Russia: Center for the monitoring and statistics of education (CMSE) of the Ministry of Education and Science. [Tsentr monitoringa i statistiki obrazovaniya, Ministerstvo obrazovania i nauki]. (2004). Retrieved March 5, 2006, from http://stat.edu.ru/stat/vis.shtml

The Slovak Republic: Source: Statistical Yearbooks of Education SR (2000, 2001, 2002, 2003, 2004, 2005).

Ukraine: Vyscha Osvita: retrospektivnyj analiz (Higher Education: Retrospective Analysis). Kyiv: Asotsiatsija Privatnyh Navchalnyh Zakladiv Ukrainy (Association of Ukrainian Private Educational Institutions), 2003; Vyscha osvita i nauka—najvazhlyvishi sfery vidpovidalnosti gromadjanskogo suspilstva ta osnova innovatsijnogo rozvytku (Higher Education and Science—the Most Important Spheres of Responsibility of Civil Society and the Ground for Innovative Development).

Kyiv: Ministerstvo osvity i nauky Ukrainy (Ministry of Education and Science of Ukraine), 2005. (Originals in Ukrainian).

Notes

1. Outside higher education, the most dramatic privatization, concentrated in a short time, has been the converting of public into private institutions, but such has almost never occurred in modern higher education. As in other places, the rise of private higher education in Central and Eastern Europe has been largely from the creation of fresh institutions. The difference in Central and Eastern Europe is that the creation of new private institutions had very few historical precedents and affected most of the countries. Compared globally, this development is late, trailed since only by the African and the Gulf-state regions. The first region to move to widespread private higher education was Latin America, which occurred by the middle of the last century (Levy 1986).
2. Most of the chapters focus on the post-communist countries of Central and Eastern Europe. Some chapters refer also to several former Soviet republics. Holmes (1997, p. 309) echoes the conventional view that the post-communist region shows great commonality, though he adds that the modern era shows greater cross-country variety than the pre-communist. The countries across Central and Eastern Europe and the former Soviet republics present a mixture of large and small countries (table I.3 in Introduction Appendix for population figures).
3. Elsewhere, as in much of Latin America, Africa, and Asia, there was a significant shift toward market dynamics, differentiation, competition, and movement away from some traditional notions of central political direction and standardization, from authoritarian regimes to relative democracy and decentralization, but noncommunist states did not have the extreme of centralization seen in communist countries. Comparable higher education and arguably economic changes came in communist countries outside Central and Eastern Europe and the former Soviet bloc, including China, Mongolia, Vietnam and Cambodia, though the overall political change was much smaller there than in Central and Eastern Europe. Moreover, in comparative context, the Central and Eastern European systems had historically limited higher education enrollment, so the new-found political and economic freedoms unleashed a huge fresh demand for higher education, one which public institutions alone could not accommodate.
4. State enterprises are playing substantially less of a role in Europe and Central Asia compared to the 1990s. By 2003, all but two countries in the region had completed or were on the verge of completing small-scale privatization, while 20 countries had privatized at least 25 percent of large enterprises . . . Yet, many countries, including EU accession countries with longstanding privatization programs, e.g. Poland, still have large "strategic" companies in competitive sectors such as steel, petrochemicals, and manufacturing that are state owned. State ownership is especially pervasive in Central Asia: in 2002, state enterprises accounted for over 50 percent of GDP in Belarus, Moldova, Tajikistan, Turkmenistan, and Uzbekistan (World Bank 2004). In the Balkans, countries such as Kosovo and Serbia have thousands of "socially-owned" firms. Residual government ownership in privatized enterprises in the region as a whole is also quite high (Lieberman 2003). And across the region, despite recent advances in

sectors such as telecoms, there is still a large stock of utilities (power and water in particular), banks, and nonbank financial institutions such as insurance companies that are still state-owned (Kikeri and Kolo 2005, p. 18).

5. Nonprofit may be defined as private initiatives "legally prohibited from earning and distributing a monetary residual" (James 1987, p. 398) or following an objective of maintaining and increasing a given aspect or value instead of maximizing profit.
6. For-profit institutions are a growing type of private higher education institutions internationally, at once very dependent upon the market along with student satisfaction and buoyed where they can gain legitimacy from public agencies, just as they are vulnerable when this legitimacy is explicitly denied.

Bibliography

Clark, B.R. 1983. *The Higher Education System: Academic Organization in Cross-national Perspective.* Los Angeles, CA: University of California Press.

DiMaggio, P. and Powell, W. 1991. The Iron Cage Revisited: Institutional Isomorphism and Collective Rationality in Organizational Fields. In Powell, W. and DiMaggio, P. (eds.), *The New Institutionalism in Organizational Analysis.* Chicago: University of Chicago Press. pp. 63–82.

Eurydice 2002. Key Data on Education in Europe 2002. European Commission. At http://www.eurydice.org/documents/cc/2002/en/CC2002_EN_chap_F.pdf#nameddest=fig_f04, accessed January 11, 2006.

Geiger, R.L. 1986. *Private Sectors in Higher Education: Structure, Function and Change in Eight Countries.* Ann Arbor: The University of Michigan Press.

——— 1991. Private Higher Education. In P.G. Altbach, (ed.), *International Higher Education: An Encyclopedia.* New York, NY: Garland Publishing, pp. 233–246.

Gupta, A. 2000. *Beyond Privatization.* London: Macmillan Press Ltd.

Holmes, Leslie. 1997. *Post-Communism: An Introduction.* Durham: Duke University Press.

James, E. 1987. The Nonprofit Sector in Comparative Perspective. In W. Powell (ed.), *The Nonprofit Sector.* New Haven and London: Yale University Press, pp. 397–416.

Kikeri, Sunita and Kolo, Aishetu Fatima. 2005. Privatization: Trends and Recent Developments. World Bank Policy Research Working Paper 3765, November 2005 at http://www-wds.worldbank.org/external/default/WDSContentServer/IW3P/IB/2005/11/08/000016406_20051108153425/Rendered/PDF/wps3765.pdf, accessed on October 2, 2006.

Levy, D.C. 1986. *Higher Education and the State in Latin America: Private Challenges to Public Dominance.* Chicago: The University of Chicago Press.

——— 2006a, forthcoming. How Private Higher Education Growth Challenges the New Institutionalism. In H. D. Meyer and B. Rowan (eds.), *The New Institutionalism in Education: Advancing Research and Policy.* Albany, NY: SUNY Press.

——— 2006b, forthcoming. The Unanticipated Explosion: Private Higher Education's Global Surge. *Comparative Education Review.* Vol. 50, May, pp. 217–240.

Lipset, S.M. 1981. *Political Man: The Social Bases of Politics.* Baltimore, Maryland: The Johns Hopkins University Press.

Meyer, J. W. and Scott, W.R. 1983. Centralization and the Legitimacy Problems of Local Government. In J.W. Meyer and W.R. Scott (eds.), *Organizational Environments: Ritual and Rationality*. Beverly Hills, CA: Sage. pp. 199–215.

Pfeffer, J. and Salancik, G.R. 1978. *The External Control of Organizations: A Resource Dependence Perspective*. New York: Harper and Row Publishers.

Suchman, M.C. 1995. Managing Legitimacy: Strategic and Institutional Approaches. *Academy of Management Review, 203*, pp. 571–610.

Tomusk, V. 2003. The War of Institutions, Episode I: The Rise, and the Rise of Private Higher Education in Eastern Europe. *Higher Education Policy*, No. 16, pp. 213–238.

Tursunkulova, B. 2005. Private Higher Education in Central Asia. *International Higher Education*, No. 38, Winter, pp. 10–11.

Part I
Regional Perspectives

CHAPTER ONE

THE LONG QUEST FOR LEGITIMACY: AN EXTENDED GAZE FROM EUROPE'S WESTERN PARTS

Guy Neave

What's in a name? That which we call a rose
By any other name would smell as sweet.
—Shakespeare Romeo and Juliette Act II Scene 2

Introduction

The soliloquy of Ann Hathaway's second husband—an inveterate scribbler—serves to remind anyone who sets out to deal with the issue of privatization in higher education that before they go very far they may have to tackle head on a distinction antique but important. That distinction involves the ancient difference between nominalism and essentialism. Indeed, when brought to bear on the notion of privatization, it is a distinction distressingly appropriate. Beneath the apparently homogenous term a veritable host of interpretations, degrees of difference, and thus understanding, abound. And, to make matters more challenging, such interpretations change shape, assume new connotations and consequences for the future profile of higher education depending on the particular disciplinary perspective, cannon or methodology one applies to its dissection. Privatization to the classical sociologist does not always carry the same significance as it does to the economist, politician, or merchant prince.[1] Add to this the basic consideration, as with everything else in comparative higher education, that the circumstances—cultural, political, social and economic—prevailing in one country merely add further nuance and variation without necessarily strengthening the overall conceptual frame. Thus, one finds oneself questioning not whether the outcome of the policy of

Privatization as it has come to be understood today and its accompanying changes in institutional status do indeed smell as sweet—for sweetness is in the nostrils of he or she who sniffs—but whether the roses one admires in their sheer variety are indeed comparable at all.

Taken at its face value rather than entering immediately into the details of how it is defined and on what criteria, Privatization as a descriptor to an overall and complex agenda stands one of the most powerful in the dimensions currently reshaping higher education as one amongst other sectors not least of which is national health services and is important on that account. At another level, privatization acts as a central force in reordering the institution of higher education to meet the perceived demands of the Knowledge Society. As such, it bids fair to be one of the new "universals" around which Knowledge Society seems to be coalescing.

Differences Historic and Systems Referential

If we are to grasp the full extent of the differences that accompany the notion of Privatization as it is currently emerging in Western Europe by contrast to its counterpart in East and Central Europe which this book analyzes with precision, sensitivity, and deftness, it is important to distinguish between privatization as a contemporary and dynamic dimension in higher education and its historic heritage. To get a more immediate purchase on the matter, like Lenin, we need to take one step back to take two steps forward. To do this, I focus uniquely and wholly on Western Europe. I do not intend to draw any analogues or to explore the ways in which Western European universities influenced other parts of the world, though obviously as the home of some of the world's referential systems of higher education during the nineteenth century—amongst which the English, French, German, Scottish, and Spanish (Neave 1998)[2]—its influence was both wide-ranging and weighty. (Levy 1986) In limiting the scope of this chapter to Western Europe alone, I do not deny—far from it—the often exceptional pattern of development that characterized Western Europe then and that today goes far in explaining what the supporters of privatization cannot fail to look upon as a certain querulousness, if not reticence to move down the straight and narrow path to salvation.

Even so, by introducing the notion of Western European "exceptionalism" in this specific domain, implicitly we open up another line of enquiry that has to do with the exact geographical locus of that normative system of higher education in contrast to which Western Europe stands precisely as an exception. This in turn leads us down the road to those long term shifts between "referential systems" as some cease to be world referents and others take their place. Seen from this angle, the shift is both obvious and explicit.

The normative and referential system of higher education that sets the benchmarking for the policy of privatization in this sector as in others—though they are not my concern here—is very clearly the United States. No student of higher education worth his/her salt will deny it, above all not when the issue of privatizing higher education occupies the center stage.

Inside the Whale

In this chapter, like Jonah, I am concerned with the view deep inside the Whale, with the "interior evolution" of higher education in Western Europe, an evolution described with reference to itself and in its own terms rather than primarily in exploring the differences with exterior systems and using *their* terminology to do so. This does not mean we can play fast and loose with those moments in history that stand as marker points when Western Europe took a different path. The "interior narrative" cannot dispense with a broader backdrop. For all that, it remains an "interior narrative" and its purpose is to dissect that long-drawn-out process of shaping the political and social constructs that brought about what may nowadays be seen as the "first modernization" of the European university and which still retains considerable weight as that institution now faces the "second modernization"—whether we happen to label the latter as the Knowledge Society, European integration or whether we examine it through the vehicle of privatization in higher education. Indeed, the overtones and perceptions, which echo across the years though varying depending on particular national circumstance, have not been without significance in shaping both the perceived standing and thus the social acceptability that attached to the notion of particular ownership. Yet today in certain cultures in Western Europe the historic association between the private, the particular, and privilege exercises a certain restraint upon advancing the policy of privatization itself and nowhere more so than in higher education. If history no longer weighs upon man's mind like an Alp, it is not entirely absent either in the present quest for legitimacy and very especially so in the university—perhaps the last historically conscious social institution to command a mass following—and one that moreover has extended its constituency spectacularly over the past three decades.

From historical perspective, the emergence of an equivalent to "private sector" of higher education in Western Europe was far less clear-cut than its counterpart in the United States, though one may note a certain similarity in the timing. The marker points in both cases are well known. The Dartmouth Judgment of 1819 in the United States set the template for university development by placing ownership firmly in the hands of trustees rather than the State (Trow 2003). By contrast, the Memorandum of Wilhelm von Humboldt on the creation of the University of Berlin (penned

13 years before (Nybom 2003)) and the establishment of the French *Université Impériale*[3] (in 1811) (Verger 1986; Charle 2004, pp. 44–45) both set the basis for a very different pattern of what anachronistically may be seen as a "developmental dynamic." Both European initiatives confirmed and opened the way for the transfer of ownership from and the incorporation of the "Academic Estate" as a self-standing guild or corporation into public service as part of the State. (Neave 2001, pp. 13–70).

Defining the Public Sector

The exact limits between the two sectors—public and private—was the subject of bitter and prolonged strife across Western Europe for the better part of the nineteenth century, fuelled in part by a drive toward replacing the *Machtstaat* by the *Rechtsstaat* (The State grounded in force being replaced by the State founded upon law); in part, by the influence of the French Revolution of 1789 that linked the modern state to the principle of meritocracy. Last, but not least, modernization also embraced the issue of how far the identity of the nation and the socialization of its administrative and political elites were to be closely associated with an institution symbolic of the Ancient Regime the Church Universal, with a national church, with many churches or with no church at all.

The Sound of Strife

Thus, what was also at stake and very explicitly so in the Latin countries as well as those where the both Catholic and Protestant communities coexisted in Germany, and the Netherlands was whether the universalism claimed by the Catholic Church in matters of education and higher education by extension should be underwritten and actively supported by the State. The alternative was for the State to set up a countervailing system of higher education with the ultimate purpose of replacing the national elite based hitherto on birth by one based on merit; an elite based on ability demonstrated irrefutably by achievement in nationally defined and nationally competitive public examinations that gave entry to higher education and which, on graduation from university attested to the possession of knowledge certified and valuated by and for public service.

These two visions of the place of the university in the political and social order were adversarial. They gave rise to bitter strife as the issue was fought out in such different arenas as the *Kulturkampf* in Bismarckian Germany, as the *Schoolstrijd* in the Netherlands, and as the *Lutte scolaire* in Belgium. Though the battle focused principally on primary and secondary schooling, the consequences naturally extended to the university world. Nor was any solution swift in coming. Only by the constitution of 1917 in the Netherlands and the so-called *Pacte scolaire* of 1958 in Belgium with similar legislation in the same year in France did the shouting and shrieking finally cease. It is then not greatly to be wondered at, though memories are often longer than politicians tend to wish, that in Western Europe today privatization, now an economic option rather than the political and religious issue it once was, should nevertheless remain an affair both fraught and delicate in the extreme.

Incorporating Higher Education into Public Service

Strictly speaking, the gradual building up of a higher education system closely aligned with public service (van Wageningen 2003) defined private sector higher education virtually as a residual function. It was a definition arrived at more by omission than by commission. For what was defined in the various Western European nations in the course of the nineteenth century was the profile of the state sector, the conditions of access to it, employment in it and what today would be termed "resourcing"—in short, its relationship with the Nation State and thus the procedures of accountability to the public via that particular relationship.

Hence, the process of laying down the ties between state and university amounted to, and at the same time was seen by those *not* included in the process of legal definition and incorporation as, a definition through exclusion. Exclusion took various forms. Amongst the most important was the *effectis civilis*, that is namely, that the degrees awarded by public universities conferred upon their holder the right to apply for public office in the national civil service. Exclusion of those establishments whose degrees did not confer the *effectis civilis* lay at the heart of what many today see as the monopoly exercised by the public university in Western Europe over appointments in the public sector. And if this monopoly proved the occasion for bitter and enduring conflict in such countries as Belgium, the Netherlands, France, Spain, and Italy throughout the nineteenth century, it remains no less a source of occasional dissent and protest today.[4]

Political and Ideological Constructs

The *effectis civilis* provided a clear operational distinction between those universities that enjoyed it and were recognized by the State from those that were not. Moreover, the latter, if acknowledged as having a right to exist, were not formally defined as public establishments. Nor were their Trustees (*pouvoirs organisateurs* in Belgian legal terminology)[5] composed in the same way. The latter were held to represent "particular" interests in contrast to the "general" interest or the interest of the Nation—literally, the "Res publica," an ideological stance that still has considerable power to rally support in such countries as France, Italy, and Spain. There, the rise of a secular State held that religious allegiance and identity were attributes of a particular group rather than corresponding to the general and public interest.

Thus, one of the consequences of the long drawn out process of incorporation of certain universities into state or public service was also to exile others to outer darkness, some of which figured amongst the most ancient of Europe's university foundations. (Neave 2001). The latter chose to retain their ancient allegiance less as private establishments so much as those not dependent on—or apart from—the State, which is far from being the same thing. Rather, as universities subscribing to a particular religious ethic or belief, as symbols of an earlier continuity and of a community that preceded the nation state, they are better described in the Western European context as belonging to a category of "nonstate," rather than private, establishments. Their trustees were "organizing powers" rather than individual owners or their nominees. Lublin in Poland, Louvain in Belgium, or the Pontifical Universidad Comillas in Spain stood as excellent examples of an institutional identity that once spanned the centuries but was now defined by exclusion on the grounds of its partiality within the nation state. Such reasoning explicitly denied their earlier claims to legitimacy that rested on the universality of the belief they upheld.

If we interpret the nineteenth-century modernization of the European university in terms of its incorporation into public and State service through the creation of an institution explicitly committed to this purpose, then by the same token we must also recognize that the residual existence of a nonstate sector that overlaps the American notion of "private" higher education is also a step in *delegitimizing*—if not stripping away—the historical legitimacy this latter group once commanded. Here the legal marker points are very clear. The Ley Moyano of 1857 in Spain, (Garcia-Garrido 1992; McNair 1984) its Italian counterpart in the Casati Law of 1862 (Martinelli 1992, p. 356), and the decision taken by the French Republic in 1875 not to support any further growth in Church sector higher education from public funds (Durand-Prinborgne 1992) were key moments in shaping the "nonstate" sector of higher education in their respective countries.

Access to Public Resources

Having the monopoly of qualifying their graduates for national administration was not however the only benefit that the title of public university bestowed. In return for being the forcing house for the cadres of the Nation, public universities were almost wholly underwritten by public finance, a benefit signally lacking in nonstate establishments.[6] The nonstate sector, its alumni, and its supporters were not slow in pointing out the paradox of the situation. If the state backed public establishments in the true name of meritocracy, equality, if not always fraternity, it also violated its own principles by refusing equality of treatment to those citizens who chose to pursue higher learning untainted by Godlessness and agnosticism. Thus, the vexed issue of access to public resources remained then, as it is today, crucial.

John Bull's Exceptional Island

To the broad thrust of formal incorporation of Europe's universities into public service, there remained one notorious exception: the United Kingdom. Far longer than most universities in Western Europe, Britain retained a degree of collegial self-government and ownership that in the two ancient English universities of Oxford and Cambridge remained intact until well into the twentieth century (Eustace 1987). It was precisely such a medieval pattern of quasi guild ownership by Fellows and Masters—often alluded to in French as the corporate interest—that had been abolished in France in 1792 (Renaut 1995). Throughout the nineteenth century, the British pattern of university development eschewed any transfer of ownership, still less the redefinition of academic staff as servants of the nation. It involved a multiplication in the modes of ownership without altering the essential feature of a system based on a nonrelationship with national administration, which effectively stood as the polar opposite to the pattern predominant in mainland Europe. The relationship between universities and government in Britain rested on the dual principle of distance from, and the nonintervention of, national administration in the affairs of academia. Individual universities were legally self-governing. Furthermore, the particular details of such self-government were laid down by individual Act of Parliament in the form of the University's Charter or founding document.

The remarkable feature of the United Kingdom's universities was not that they remained private in nature and governance but rather no attempt was made to incorporate them as state establishments. Even their funding from national taxation revenue, introduced in 1919 together with the creation of the University Grants Committee (UGC) that oversaw the sharing out of public subsidy to the universities, remained untainted by the notion of incorporation. However, recently, some have argued that the unintended consequence of this arrangement served in the long term to place Britain's universities in a position of financial dependence on central government, a situation which, if not incorporated *stricto sensu*, was not far removed from it (Scott 2000). The British economist of higher education, Gareth Williams has pointed out that such a funding pattern bore more resemblance to a philanthropic arrangement than to an instrument of public oversight and control (Williams 1992, pp. 69–95). Under such a covenant the UGC (abolished in 1986) acted on behalf of all universities. It negotiated directly with the Treasury (Ministry of Finance) without passing through the intermediary of the Ministry of Education. In turn, the overall Treasury allocation was distributed to individual universities on a lump sum basis by the UGC that served as a buffer against further government incursion—a self-denying ordinance[7] expressly conceived to uphold the legitimacy of distance between government and university.

Contrary Imaginings

What can be said about privatization in its historic form in Western Europe? The first and very obvious feature is that with the possible exception of the United Kingdom, defining institutional identity and status in terms of privatization played little part. Rather, the task of definition turned around setting out the status, financing of universities, and conditions of academic work, laying down the boundaries of the state university system as a national undertaking. The second and equally obvious characteristic is the corresponding shift in legitimacy away from "nonstate" establishments above all in the major centralized states of Western Europe—France, Italy, and Spain. Legitimacy attached to public service rather than to serving interests thought to be redolent with long-established social privilege. And whilst the notion of social efficiency through individual mobility and competition could most assuredly be accommodated within the revolutionary slogan of the "career open to talent," such competitive virtues were thought to reside in state sector universities, and rather less in their nonstate rivals that carried with them the burden of being associated with a political order perceived as both traditional and deeply hostile to the then contemporary version of modernity.

Yet, the notion that Western Europe fulfilled a function roughly similar to privatization did not bring with it the counterlegitimacy which is deeply embedded in nineteenth-century British and American liberalism—namely, that the state should be kept at a distance so that individuals and communities could flourish the better. In the United Kingdom, such a nineteenth-century liberal construct added an important dimension to reinforcing the legacy of academic guild self-government by grafting on to it the liberal ideology of governmental nonintervention in British universities. However, such a liberal add-on did not serve to distinguish one university from another or to segment particular types of university. Rather it applied to universities in general and thus served to preserve them as self-governing institutions.

There remains, however, one final point. It is central in understanding the difference between (*å l'américaine*) public and private universities, public and private *å l'américaine,* and its Western European counterpart. In effect, even those universities in Western Europe that, seen from American eyes, bear a degree of similarity in status to "private universities" never regarded themselves in this light. To have accepted the qualification "private" would have been tantamount to accepting as legitimate the relative marginality that incorporating other universities into state service assigned to the "non-incorporated" sector. For the latter to have done so would have equally been to recognize the very status that others wished the "nonstate" sector to have. Marginalized and exiled, though they were nonstate and denominational establishments did not give up their claim to serve the interests of part of the collectivity and the collective identity that went with it—Catholic, Protestant, Free, or Pontifical. Indeed, this burning conviction goes far in explaining the bitterness that lay behind the battle for the public groat. And whilst it has to be admitted that it is equally unlikely "confessional" and above all, Catholic establishments would have accepted the distinction of "nonstate" that we have used to identify them, it is no less clear that describing them as "private" is a misnomer in the setting of Western Europe. In their own eyes, confessional universities were both *of* their community and very much *in* it.

Change and Continuity

If we have taken a little time to set out the historic background to what some may care to see as Western Europe's homegrown equivalent of privatization, we have done so because the values and attitudes that accompanied it are not without relevance when it comes to dealing with the present day imported variety of the same. Differences in perception, status, standing, and legitimacy of the nonstate sector in Western Europe mark it off from the United States. They also split off Western Europe from those nations to the East that, in the aftermath of war, found themselves having to swallow a model of higher education together with its attendant political, economic, and social values that arrived in the baggage train of the Soviet army[8] (Neave 2003). The historic form of privatization *å l'européenne* as a residual function to the greater task of incorporating higher education as a public service, did not, however, cease with the end of the war. On the contrary, an excellent argument can be made for seeing the post-war period in Western Europe as a speeding up in the further construction of the public sector, adding spectacularly to the institutional infrastructure, and doing so through a slightly different mode of justification from that which attended nineteenth-century nationbuilding and socialization of political and administrative elites.

If justified as a continuation of a social policy that extended the principles of the welfare state to higher education, the massification of higher education remained fully within the European practice of collective rather than private initiatives. Indeed, had the private interests of employers been taken fully into account at the time, there is good reason for believing that mass higher education would never have taken place at all. Even so, massification was borne aloft by public universities. Not surprisingly, the numbers of public universities multiplied accordingly (Neave forthcoming). Significantly, it also brought about a lowering of barriers between public and nonstate higher education and did so in Belgium, France, and the Netherlands by opening public funding to the latter (Geiger 1986; Frijhoff 1992). Massification, however, was wholly a state initiative in Western Europe and one to which the State remained committed, even in the teeth of student unrest that spread from Paris, Berlin outwards to Stockholm and Amsterdam in the aftermath of May 1968.

Massification

Massification contained the seeds of its own discomfiture, which took root in two forms. The first sprang from the very success of massification itself. By the mid-seventies, the numbers of those graduating from Europe's mass universities exceeded the ability of the public sector of the economy to absorb them. The second involved levels of student funding and subsidization that had accumulated around higher education when higher education was the affair of elites. With expansion, established patterns of student and institutional finance were rapidly perceived as ruinous when extended to the mass—a situation made worse by the very success of national policy in meeting social demand. In short, the ability of the public purse to bear the cost could no longer be counted upon. The upshot was to redefine the notion of privatization, though not entirely separating it from the historic associations and *sous entendus* from an earlier age that still attached to the term. From this perspective, privatization in Western Europe moved on from the issue of possession, ownership, and social reproduction to embrace a wider and infinitely more complex agenda, namely, how to recouple the university with the private sector of the labor market as the major outlet for its graduates. Thus, what had once been construed as a political issue rapidly moved over to become a central item in economic policy.

Paradox, Ideology, and Pragmatism

Here was a situation rich in paradox. The paradox that permeates the Western European version of privatization is not that it ostensibly bids fair to unravel a model of development that focused on the public sector for the best part of a century or more, though the point is well worth the making. The paradox resides in the fact that whilst the process of privatization in Eastern and Central Europe emerged from the collapse—moral, political and financial—of central state administration and the bankrupt alliance of State and Party that held it in thrall, in Western Europe the equivalent of privatization required the intervention of the State to bring it about. Thus, in Western Europe, privatization has a dual face. It does not start off from the basic premise, shared as much by Britain as by the United States, that collectivity can flourish only when individual initiative and competition are given full run. These ideological overtones are not absent, but tend to be less virulent in their challenge to the collectivity as expressed through the State. One can detect a certain similarity in the pragmatic measures taken to generate additional resources and to sustain the demand for higher education that in certain systems has gone beyond 40 percent of the age group, which Martin Trow identified as the stage beyond mass higher education—universal higher education—which others are now setting themselves as a nonnegotiable ambition (Kwikkers et al 2005).

In effect, for the past 15 years Western Europe has been concerned with reform in the financing of higher education, with cost-sharing (Johnstone 2004) with the generation of institutional revenues from sources other than the public purse, with the prospect—sometimes near, at others rather more distant—of full-cost student fees. All of this is a significant redefinition from the original connotation this term once carried with it, namely, the notion of the ownership and particular identity of individual establishments of higher education.

Such are the operational—one is tempted to say, pragmatic—aspects of what is in essence, part of a broader shift in the purpose of higher education.[9] This shift in the purpose of higher education has turned its mission from modernization construed as a political function toward acting as a central instrument for economic overhaul, a major watershed in which politics serves to define the choice between different economic options, their implementation and their selling to a bemused citizenry, in effect negotiating the terms of their social acceptability which is one stage prior to acquiring legitimacy (Neave 2004).

Two Roads Toward Privatization

Against this backdrop, the notion of a wholesale transfer of ownership and the rise of institutions in the hands of individuals or corporations overwhelmingly reliant on student fees or on contractual services, which one sees in certain systems in the Central and Eastern parts of the European landmass—notably Rumania, Poland, Russia *entre autres*—shines in

Western Europe by its low profile. Still, privatization is present, though there is an important distinction to be drawn between new establishments, created de novo and claiming a private status and those that, until recently an integral part of the public sector, decide to opt out and seek their fortunes in relative independence from public funding and from what some feel to be the constraints that national legislation imposes.

As an illustration of the first, one may cite the opening of private law schools in Hamburg Germany, the founding of private universities on the banks of the River Oder, the creation of so-called University Colleges as a variant upon the American liberal arts college, usually associated with revenue raising by some well-established universities in the Netherlands and Belgium. Clearly, the appeal of privatization in the strict sense of redefining ownership exercises a certain charm. The second variation is no less interesting because those taking up this option tend to be universities that have already built up an enviable reputation as public sector establishments and, as a result already command a solid financial relationship with the private sector. Hence they are less reliant on public resourcing and reckon that even greater advantage is to be had by changing their legal status. Amongst such fortunates in Sweden, for instance are Chalmers Technological University in Gothenburg, the University College at Jönkoping, the London School of Economics and Political Science, and Warwick University in the United Kingdom. Equally interesting is that their legitimacy derived not from their activities as private establishments, which they already possessed. The fact of their opting out of the public system strengthens the credibility of privatization as a possible alternative.

Nevertheless, "going private" at the moment does not appear as a possibility open to all. With the exception of the Portuguese Polytechnics that have not flourished (Amaral and Teixeira 2000, pp. 245–266) those that have embraced privatization most eagerly in Western Europe tend, by and large, to be elite establishments with strategically central specialties, high research capacity and close ties to key sectors in the post-industrial economy—Engineering, Information Technology, Banking, and Business Administration. Such individual cases stand more as exceptions that prove the general rule. The general rule, so far, remains that whilst new and alternative forms of ownership are putting their heads above the parapet, the general thrust is slower, more cautious, and in such systems as France and Germany, a matter of far greater political delicacy than has been the case in the lands to the East.

Explaining the Pace of Change

Yet, in all this, one deceptively simple question clamors for an answer. Given the historic marginality of nonstate higher education in Western Europe, given too that higher education reform has long been national in scope, how can we explain the speed with which the *idea* of privatization established itself, both as an underpinning to a program of reform and as an ideology with the potential for "reconstructing" higher education in Western Europe? It is a question worth posing if only for the fact that those in favor of it, as well as those for whom privatization *is* part of their particular landscape in higher education, tend to take its extension elsewhere for granted as a natural and logical outcome and a triumph for that same rationality which underpins *their* heritage. Still, when we bear in mind the power of collective action in Western Europe and that massification reinforced the power of central government, that too an ideology so fundamentally at variance should make headway and that so rapidly deserves to be explained rather than being simply noted *en passant*.

The first point that deserves our attention is that the measures which, with hindsight, may now be seen as a break from what has sometimes been termed the "Keynesian consensus" in planning higher education development, were not perceived as such at the time. The budget cuts that spread across Western Europe beginning with the United Kingdom in 1981, affecting Belgium by 1986, and the policy to concentrate and rationalize higher education resources in the Netherlands, were not identified either as the forerunners of system-wide overhaul and still less as a shift in political ideology from the collective to the individualistic. Rather, they were viewed as part of that natural husbandry, the pragmatic action that governments sometimes have to take. In other words, privatization as a way of perceiving higher education's development in the future was far from evident as a solution that had promise. Rather, privatization followed in the wake of initiatives dictated by administrative pragmatism. As many observers have pointed out, the sequence of pragmatism preceding ideology was apparent even in Britain—the first system in Western Europe to explicitly conjoin higher education policy to the notion of privatization (Williams 2004).

To raise pragmatism to the heights of an ideology is not unusual. However, it poses a further question: why lay on an ideological gloss at all? In part, the answer is to be found in the nature of the measures Western European governments envisaged—and more to the point, introduced—to meet the crisis in public expenditure. They were a radical departure from established practice and went in the face of the whole thrust of higher education policy over the preceding two decades. The long-drawn process of incorporation and privatization was, in truth, radical in root and branch. By the late-eighties, it became evident to governments in Britain, Belgium, and the Netherlands that changes in the economy could not be handled by piecemeal tinkering alone and very certainly not in an institution as central to the Knowledge Economy as higher education (Williams 2004, pp. 241–269).

Short-term pragmatism yielded before the prospect of a sustained overhaul. In such circumstances, to point to examples elsewhere of policies similar to those one envisages and very especially those that have the advantage of having been tried, tested over time and, above all, proven successful, is no small benefit.

Privatization as an Ideological Accelerator

Privatization has three principle components to it: as a policy, as a process, and as an ideology. In the latter setting it may be seen as the polar opposite in its subtending values to what was described earlier as "incorporation." The latter held that modernization is best secured by collective effort and rationalization through regulation, law, administration and government. The former takes the contrary view—namely, that virtue lies in deregulation, individual boldness, initiative, and competition, between the privately possessed. Similarly different are the notions each holds about privilege, its perpetuation, and its location. Whereas incorporation saw privilege in historic terms, of social classes and their identifying beliefs, and sought through higher education to contain them by limiting them to the private or nonstate sector, privatization stands privilege on its head. As a doctrine of belief and social development, privatization redefines privilege primarily as those institutional and historic features that it identifies as obstacles to economic efficiency and very particularly the dysfunctional protectiveness that overmighty governments exercised over public or state universities. Amongst the boons, ever to be regretted, that resulted from such protection are guaranteed employment for individuals and historic incrementalism in financing the institution. Both are held up as irreconcilable obstacles to the fundamental driving force of business and social development, to wit individualism and competition. Thus privilege is now identified with public universities. Competitive derring-do dwells in the nonstate sector of the higher education system, just as it is held that in the future, well-being in the Knowledge Society will be determined by the performance of the private sector of the economy.

As a policy, privatization is neither exclusively economic nor financial in its consequences, though the discourse that drives it today very certainly is. It is rather a "many splendored thing." Beginning with the classic definition around ownership, the legal status of institutions and the grounding of their obligations and responsibilities sometimes in Constitutional, Administrative, Public or Private law *pace* in Western Europe such matters as conditions of student access, enrolment charges, conditions of employment of academic and other staff, their pay scales (often subject to nation-wide legislation) (de Weert and van Thyssen 1998) has moved on to embrace institutional efficiency, performance and accountability, responsiveness, accommodation to student demand and so on. It is a complex and wide-ranging agenda. It is also a species of conceptual *omnium gatherum* if only for the plethora of goals, purposes, initiatives, changes, and enactments that are associated with the term. The sheer variety in the measures that may be associated with the process of privatization—that is, measures associated with becoming private or reforms legislative, legal, institutional, and financial that provide pointers to a particular system's progress along the path toward this desirable state—are equally striking. The very variety of initiatives associated with privatization explains in part the speed at which privatization as a mobilizing construct has spread across the face of Europe. It explains it, however, often as an artifact of definition.

The Definition as Artifact and Politics

It does not follow that the privatization introduced in one system of higher education corresponds to—still less involves—the same dimensions in another. Nor does the use of the same generic term imply the combination and range of initiatives being addressed are the same as those being tackled by others. But, the wrapping up of very different measures and their labeling as privatization gives the impression of a certain convergence simply by dint of employing the same term to bundle them together. Governments, like gastronomes in a Chinese restaurant, may choose from the menu of privatization and may indeed qualify their policies as such. Each tidbit is tasty and figures somewhere on the menu *a la chinoise*! What the individual diner puts in his rice bowl, however, is very different.

From this we may conclude that convergence, which rests on a shared name, does not essentially mean similarity, since, it may also cover major differences. Still, it does fulfill a most important political function. It bolsters the impression of rapid dissemination, which in turn feeds the impression of an almost juggernaut inevitability—of progress to be ignored at one's peril, a species of domino effect in reverse. The claim often made by governments that initiatives taken at home bear a broad similarity to those endorsed by others abroad, has of course a purpose: to bring about greater domestic acceptability and credibility for the measures that government chooses to identify and equate with privatization. By demonstrating the legitimacy and success such policies have had abroad governments seek to bedeck their own policies with a similar fig-leaf! The radicalism of such measures is thus offset by the necessity of their advancement and all the more so because others are already moving in the same direction. International comparison is then a most useful tactic in seeking to acquire legitimacy for policies that, on their home ground, are nothing if not highly controversial. And in politics—as in wooing—speed is very often of the essence. Even so, whilst the descriptor that the governments may set upon their program may indeed be the same, the agenda is very different.

Degrees, Differences, Presence, Absence and Quibblings

What such ambiguity also tells us is that there are degrees of privateness depending on the comprehensiveness and the range of measures proposed or introduced. From a strictly linguistic standpoint, privatization is less a goal so much as the process by which that goal is achieved. It is, literally, the way of becoming private. In effect, the particular items included in the process can range from a single dimension—say financial diversification—through to the wholesale comprehensive transfer of institutions from the constraints of public law to the delights of private legislation and beyond in combination with new sources of revenue and support, overhauling structures of governance, and so on. Each of these elements involves privatization to some degree. But does the fact of one dimension being privatized mean that the system in which it takes place, is for that reason now to be deemed private?

Such quibbling in its turn raises issues of the utmost nicety. Amongst them, whether the degree of privatization introduced has any significant consequence in changing the status of the higher education system in general. It also begs the question when and on what criteria privatization can be said to have gone beyond becoming and is achieved? Simply by the presence of a few institutions calling themselves private? Is there, for instance, a threshold in the fulfillment of certain criteria beyond which it can reasonably be said that institutions or systems have ceased privatizing and have become private? Consider the issue of financial diversification. Does the fact that individual institutions are less reliant on public expenditure than they were constitute privatization? When compared to a previous age, the answer is very often "Yes." And accordingly that establishment or system, depending on the level of analysis, is very often held up as illustrative of privatization achieved.

Nevertheless, such a degree of *relative* privatization, tends to pass rapidly over the reverse of this particular medal—namely, that public expenditure may often still remain the largest single revenue source of the establishment or, for that matter, for the institutional sector as a whole. Is there then a threshold, a tipping point in the process of privatization, at which the balance between the nature of a particular system of higher education ceases to be public and may be said to be on the last furlong toward privatization fulfilled? Is there any general agreement on which criterion or set of criteria that allow us to make this claim? Can one argue, for instance, in the domain of budgets, that an institution has ceased being public once it depends on public expenditure for 50 percent or less of its annual income? Or should the tipping point be placed lower—40 or 20 percent? This is a key question that distinguished scholars have raised in other settings and that in an earlier age! (Levy 1986) Yet, such central and operational criteria are necessary if the understanding we wish to have of this policy is to be grounded in reality rather than in caricature or in the realm of the symbolic.

The Binary Illusion

Elements such as these—whether privatization is to be seen as a budgetary phenomenon, as a percentage of students in the nonstate sector, not to mention their operationalization and their attendant nuances—are crucial in determining how far a system has gone along the straight and narrow way that leads to a *privatization salvatrice*. Yet, the way in which privatization is often presented fails all too often to take this multidimensionality—let alone the dynamic it implies—into full account. On the contrary, very often the perspective employed is a binary one—that is to say, whether there are elements of privatization present or absent from an individual system. In short, the plotting of privatization onto the landscape of higher education on a comparative basis is reminiscent of medieval cartography, flat and without the vital features that three dimensional topography reveals, namely difference between the heights of summits and the depths of valleys as opposed to their mere presence or absence. Presenting privatization simply in terms of its presence as vouchsafed by government intent, in measures, legislation, or changes in practice rather than seeking to weigh them and compare their significance both within systems across time and between systems at a given moment tends to leave the impression that irrespective of whether it is a policy of intent, a policy in the making, or a policy fulfilled, privatization is a notion that may appear more weighty than grounded practice would warrant.

There is, however, another angle of approach to this general issue and that is when privatization is construed in a precise and limited way in terms of reduction in public expenditure, in other words, privatization viewed as cuts in the higher education budget, or its concomitant, the diversification of funding sources and the scrabble of institutions to make good the cuts by seeking alternative sources of finance through contracts for research, development and services with the private sector. No one will disagree that this trend has been immensely powerful in altering institutional behavior, changing patterns of authority within universities, and changing the relationship between state and higher education.

Finally, though we argue later that there is another perspective and interpretation which, in Western Europe runs in parallel to policies that in varying degrees tend toward privatization, it is important to recognize the symbolic importance of contemporary Privatization. From a historical standpoint, privatization stands as a species of match replay of that seminal traffic in ideas and individuals, which in the late-nineteenth century transposed the German research university into its American variant (Ben David 1978 Clark 1994; Gellert 1997). Privatization reflects the place of the United States as a world referential system. More to the point, it also reflects, to a very considerable extent, an attempt to reproduce the conditions that favored its rise to preeminence (Neave 1998). Success in the form of the American Graduate School, of the Research University, and privatization as a way of replicating that success provide pragmatic examples of the way many governments feel higher education ought to be organized. Indeed, much of the legitimacy the construct of privatization enjoys rests upon the hope that similar success will follow on the establishment in Western Europe of broadly similar ways of doing things. Like most hopes, it combines no little millennialism with a dash of pragmatism extrapolated.

The Judiciousness of One Term

Nevertheless, what tends to be gathered together in Western Europe under the rubric of privatization, often for reasons of political convenience, opportunity, or in an attempt to demonstrate the viability of the concept to win over hearts and minds to the principle itself, may also be described with equal felicity in terms of the transfer of responsibility *away* from national administration. In view of the primordial role government has played in defining the profile of higher education in Western Europe these two hundred years past, arguably it is no less appropriate. Hence, the central dynamic of privatization may just as well be seen as an exercise interpreted as offloading, or as the devolution of responsibilities previously vested in national central administration. In other words, in the setting of Western Europe the current reality of privatization resides in its "rolling back the frontiers of the State" rather than as a root and branch reconstruction. This is not to say that root and branch reconstruction is absent. But its justification and rationale, both rest upon a rather different instrument of mobilization.

. . . and the Appropriateness of Another

When one views matters from this perspective, however, a powerful and alternative agenda emerges in Western Europe. It is no less accommodating to the notion of privatization, though it does not wholly subsume it. Still, most of the pragmatic measures equated with privatization can just as well be grouped together under a second, alternative construct. It is, however, a construct very different from privatization though this latter may indeed be accommodated within it.

If we lay aside for a moment our use of privatization as a descriptor or as a label to a complex program and concentrate rather on the change in the *site* and location of functions and tasks this process entails within higher education—some of which have been mentioned—it is perfectly possible to place the transfer of responsibility into a framework very different from privatization conceived as the Ark of Covenant by neo-liberalism. The alternative construction, which in Western Europe may be seen as a conceptual Tetrapak to contain the pure and pasteurized milk of privatization, involves a rather broader concept and one that antedates the writings of Adam Smith by almost half a millennium. This alternate theory is important in several respects. It is the notion of subsidiarity (de Groof, 1994a).

Subsidiarity and Privatization

Subsidiarity is an elegant notion, originally developed by the medieval schoolman, Thomas Aquinas (1225–1274) in the *Summa Theologica*. The principle of subsidiarity states that decisions should be taken at the level where their implementation is most effective. Thus, it supports the principle of the devolution of responsibility. However, it does not rule out its opposite, namely concentration or coordination, if need be, at a higher level than has hitherto been the case. In effect, this same principle underwrites regionalization, devolution, or the repatriation of functions back to the institutional level (Neave 2001). Equally, however, it may drive in the opposite direction, toward reinforcing coordination at the supranational level.

There are several excellent reasons for seeing subsidiarity as a weightier and more appropriate overarching theory in the setting of higher education policy as it is shaping up in Western Europe than privatization. This is not to rule out the fact that changes in funding, governance, levels of responsibility hived off from national administration and laid upon, as well as the tasks assumed by, the individual university cannot be handled within the framework of privatization. Rather, it is to suggest that a broader theory exists in Western Europe, one that is a major guiding concept in Europe's current redefinition of its self-identity. It is also a higher-level theory. Subsidiarity as a theory of administrative and institutional rationalization pursued simultaneously across national jurisdictions can both accommodate Privatization and, at the same time, also serve to explain other developments that, when viewed uniquely within confines of the latter, would otherwise appear contradictory if not downright irreconcilable.

A Modest Suggestion

Let me suggest that privatization as an economic program brought to bear on higher education may indeed be international in the sense that one sees its application across many countries. Its basic presumption, however, is that the Nation State or its variants stands as the highest level of administrative aggregation and decision-making. From this viewpoint, privatization is not international. Agreed, it is present in many different nations and those contributing to the present study make this plain. But whilst privatization shares certain features of subsidiarity—repatriating certain functions back to institutional level being one—it is largely unidirectional. It works downward with the idea that efficiency and effectiveness result when institutions—just as individuals—are masters of their own fate. It is also a policy applied within the Nation State. It does not seek to develop institutions and levels for its application beyond the Nation State—or for that matter between them—though some might argue that the former task is precisely what the General Agreement on Tariffs and Services is designed to fulfill. Seen in the context of European integration, privatization is, in effect, a program designed for national application within the individual State. It has perforce to adapt itself to the differing circumstances that prevail in the individual State. It has not—or at least, has not yet—officially acquired the status of an intergovernmental program and most certainly not the quality of a supranational policy with its attendant cross-national agencies, mandates, and their *ententes*.

Yet, it is precisely the emergence of a level of decision-making beyond the Nation State and across them that constitutes the main thrust—and for that matter the main challenge that Europe—and its higher education systems as a way of bringing that about—both face. Privatization seen as coterminous with the diversification of resource acquisition, revenue diversification, and the reimposition of full costs back onto the student and the family, certainly figures as a subagenda. This instrument is far more fragile than many believe, if it does not have the dangerous potential of turning against the hand that wields it. It is more fragile because, with few exceptions in Western Europe, student fees are rarely full-cost, though the principle of fees is now generally admitted. Still, with the exception of the United Kingdom at undergraduate level, few Western European governments have yet gone very far down this road.

The Other Face of Subsidiarity

Commitment to the other face of subsidiarity—that is, in effect, to devolution upward—progresses no less, though it has to be admitted that devolution of functions downward—the regionalization of higher education funding in the United Kingdom, Spain, Belgium, France, for instance—is there for all to see. *De-étatisation*, which is a feature subsidiarity partially shares with privatization, does not lie uniquely in devolution downward. On the contrary, the mere existence of the European Higher Education Area implies a mobility of functions both below the Nation State and the emergence of others above it. The recent proposal for a European Research Council, put forward by the German government, is more than a straw in the wind. At another level, the move amongst recently developed agencies of oversight in such domains as quality assurance and accreditation to intensify interagency consultation under the auspices of the European Network of Quality Assurance Agencies (ENQAA) established at Helsinki in 2001, is another (Schwartz-Hahn and Westerheijden 2004). It shows that the specter of a cross-national layer of understanding—and understanding almost always in such circumstances is but the first step along the path toward formalization, the acquisition of influence, the devising of procedures, and eventually the possession of authority—has gone beyond the stage of a pipe dream.

Enacting the principles of subsidiarity, above all in the area of *de-étatisation* upward, poses some very uncomfortable questions for the policy of privatization. To begin with, one of the credos in the theory of privatization is deregulation, on the principle that bureaucratic heavy-fistedness is death to initiative, innovation, and creativity—the essential commodities of the university, just as they are the lifeblood of the Knowledge Society. Yet, privatization in the sense of enhanced institutional latitude to raise money from services sold, facilities hired out, and contacts tendered has gone hand in fist with an astounding multiplication in the numbers of agencies of oversight as well as the reinforcement and steering capacity of those that earlier fulfilled a facilitatory function, not least amongst them national research councils and funding bodies. Relocating control and oversight from input to output, from per capita funding to performance, objective or criteria-related financing, their detachment from Ministries—of Education, Education and Skills Universities, Technology and Research—and their relocation in single-purpose para-statal agencies of assessment, evaluation, and accreditation has been the Leitmotif in the higher education policy of Western Europe for the past 15 years or more.

Contradictions Apparent

It is, however, in no way a diminution in the degree of control, irrespective of whether that control or the watchful eye are labeled and justified in the name of accountability, efficiency, or value for money. Control remains control and regulation, regulation regardless of whether its myrmidons sit at the seat that dispenses largesse or grace the desk that checks the accounts—and the performance—afterward. Control that governments apparently yielded earlier is replaced by another version of the same exercise in the name of the marketplace—by governments acting in that strange capacity which economists sometimes qualify as a "pseudo-market." Thus, a strong argument can be made that the instrumentality such agencies wield is if anything more pervasive and certainly endowed with far greater and immediate consequence for causing happiness or inflicting woe upon the Senates and Aulae Magnae of Western Europe's universities than was ever the case with those ministries whose "strategic oversight" they now supplement. With such systems of audit and public expectation, it requires the nicest of judgment to determine how far a control reinforced is, if at all, offset by institutional latitude. Or, for that matter, whether institutional latitude, which may be seen as one of the benefits of privatization, has any substance that is worth the having.

Therefore one cannot avoid the feeling that despite the rhetoric of privatization, the reality of higher education policy in Western Europe appears to be sadly contradictory when viewed solely within the canons of that theory. How is one to reconcile, for instance, greater institutional latitude with the growing weight of an evaluative state, which with very little imagination and still less further initiative, is on the point of transforming itself into a key element in a supraordinate sphere of coordination and norm generation? (Neave 2004a). The answer is, of course, that developments that appear to stand at loggerheads with the precepts set out in one theory are perfectly reconcilable when one changes the theory. Or that one revises the scope of that theory to make it an element—a subset—of one broader ranging and more inclusive.

Envoi

Assuming one accepts the argument that in Western Europe privatization occupies the position of a subagenda within the broader reaches of subsidiarity, the question that follows must surely be "What has been the contribution of privatization to advancing higher education?" To deny its importance would be folly or, worse still, bad faith. For, most assuredly, privatization as a mobilizing idea has played its part. It has done so in that most subtle and difficult of domains—how institutions and their workings are perceived. In this, it has altered vocabulary, institutional and system metaphors, and not least, that which is best presented as the basic referential model against which institutional progress is compared, weighed up, and judged (Neave 2005b). It has shifted higher education's basic referential institution from the State and from the referential practices grounded in the civil service of the Nation to those of the firm or, to be more precise, the industrial conglomerate since universities, unlike ordinary firms, are not single product undertakings. When one considers that the State has been the basic referential model—in many cases, for the best part of a century and a half, when one considers that this relationship has existed the longest in Europe for it was in Europe that the Nation State was forged, this is no small achievement.

What remains unclear at the present moment is whether privatization will go beyond acting as the mobilizing idea, the vision that puts the established and the acquired to the question. And, more to the point, what are those particular conditions that enable it to do so? This is an essential task and one to which the contributors to this analysis tackle with vim and determination. In doing so, they remind us that, even in a world where nominally similar agendas are unfolding with a speed that is astounding, there is always diversity and difference and both demand close and unwavering attention, just as they require a subtle sensitivity if the example and experience of others is to serve us well.

Notes

1. For the classical Greek philosophers the connotations of privacy were anything but positive. Indeed, the term used to describe the private individual as opposed to the active citizen participant in the affairs of the πολιζ—the collectivity—carried pejorative overtones. Such an individual was qualified as ιδιψτεσ which shows as plain as ever one might wish that economics and ideology can accomplish even the most radical of semantic shifts!
2. Some proponents of the notion that the United Kingdom was a single system may balk at the distinction between English and Scottish. Invidious though it might be, this distinction holds, especially in the cases of Australia, New Zealand and, strangely, Chile (Jaksic 1989, pp. 22–24). The obvious case where the practices of both Nations influenced those of a third is, of course, the United States where, if some of the more ancient East Coast establishments drew inspiration from the University of Cambridge, others further in the interior drew theirs from the Universities of Edinburgh and Glasgow (Rothblatt 1998). The German influence in the form of the Research University, first introduced with the foundation of Johns Hopkins University (Gellert 1992) serves to underline the general point whilst preserving it from accusations of parochialism and partiality.
3. To be more precise, the *Université impériale* was a species of *faux ami* since it involved far more than just the university. Rather it embraced the whole education system from primary school through to higher education. Under this arrangement, the university level bore a strange similarity to what today is sometimes presented as a "network university" with outposts—otherwise known as Faculties—in the largest cities. Universities as separate entities, abolished in 1792, were not recreated until 1896 (Durand-Prinborgne 1992; Weisz 1983).
4. Exclusion from public administration on grounds of belief—or its absence—was not an innovation of nineteenth century French Republicans or for that matter of Belgian, Dutch, Italian or Spanish democrats, seeking to contain the ravages of inherited privilege by opening up the educational pathways to a career open

to talent—far from it. It had an ancient and very respectable lineage that antedated even the university itself. Thus, one of the earliest instances on record of public rectitude enforced, occurred in fourth century Constantinople, when Emperor Theodosios I decreed that pagans were not eligible to hold public office! (Herrin 2006).

5. For this see de Groof, Neave and Svec (1998).
6. The way this difference was operationalized is interesting in itself. In France, Belgium and Germany for instance, it took the form of two distinct types of first degrees: those conferring eligibility to apply for posts in the national civil service, termed *diplômes nationaux, grades légaux* and the *Staatsexamen* respectively. Alongside them were *diplômes de l'université* in France and in Belgium, the so-called *grades scientifiques.* In the case of France and Belgium, public funding extended only to the former and the recognized programs that lead up to them. And whilst universities were certainly free to award their own degrees, these possessed neither the official status and still less the very real value on the labor market of state recognized diplomas.

   Variations on this distinction were also to be seen in Italy and Spain, where state recognized diplomas were the monopoly of public universities in the first. The explicit linkage between public service and the public university in the case of the second operated slightly differently, less through the degree awarded so much as recognition of the curricular content or "pathway" that lead to eligibility for employment in public service, a specificity contained in the term "carreras" which, as the name implies, prepared the way for careers in national administration.
7. For *aficionados* and the curious, the Self Denying Ordinance was passed by the English Parliament on April 3, 1645. It stipulated that in time of war, no Member of Parliament could hold military office or for that matter any other office appointed by Parliament. Its purpose was to remove certain aristocratic Generals in the ranks of the Parliamentarians, who were somewhat reluctant to inflict defeat on the King, Charles I, then busily engaged in waging civil war with his subjects. By extension, a Self Denying Ordinance is a situation in which an individual or body, having the right and capacity to do something, decides unilaterally not to make use of it and thus voluntarily imposes a restraint on its own legitimate and recognized powers.
8. Indelicate though it might be to suggest it, the emergence of national variations upon the Soviet model of higher education in Central and Eastern Europe can just as well be interpreted as yet another version of the historic process of incorporation, though obviously *à contre-coeur.*
9. Beneath the notion of pragmatism lurk issues delicate indeed. For pragmatism can itself be driven either by sheer necessity or stand as a choice, the latter largely being a function of political foresight, the former imposed by the absence of the same. Such a distinction, useful though it is, ought only to apply to national levels of decision making for once one penetrates down the chain of implementation, the choice of those farther up tends to become the necessity of those lower down and as March and Cohen (1974) have pointed out, the converse may also emerge at institutional level when the necessity as defined by governments is reinterpreted at institutional level to become a matter of choice. It is, of course, this distorting effect that has caused not a few governments in Western Europe,

the most notorious being the British, whose pragmatism chosen is enforced by national instruments and agencies of compliance, and thus became and remains a matter of institutional necessity, not least by the threat of performance-related budgeting (cf Neave 2005c).

Bibliography

Amaral, Alberto and Teixeira, Pedro. 2000. The Rise and Fall of the Private Sector in Portuguese Higher Education. *Higher Education Policy*, Vol. 13, No. 3, September, pp. 245– 266.

Ben-David, Joseph. 1978. *Centers of Learning: Britain, France, Germany, USA*. New York: McGraw Hill for Carnegie Commission on Higher Education.

Blom, J.H.C. 1980. De tweede wereldoorlog en de Nederlandse samenleving: continuiteit en verandering. in C.B. Wels et al., (eds.), Vaderlands Verleden in Veelvoud. deel II: 19e-20e eeuw. Den Haag: Martinus Nijhoff.

Blückert, Kjell, Neave, Guy and Nybom, Thorsten (eds.), (in press). *The European Research University—an Historical Parenthesis?* New York: Palgrave.

Charle, Christophe. 2004. Patterns. in Walter Ruegg (ed.), *A History of the University in Europe*. Cambridge: University Press. Vol. III *Universities in the Nineteenth and Early Twentieth Centuries (1800–1945)*, pp. 33–75.

Clark, Burton R. 1994. *Places of Inquiry: Research and Advanced Education in Modern Universities*. Berkeley, Los Angeles, London: University of California Press.

Cohen, M.D. and March, J.G. 1974. *Leadership and Ambiguity: The American College President*. New York, NY: McGraw-Hill.

De Groof, Jan 1994a. *Subsidiarity and Education: Aspects of Comparative Educational Law* Antwerpen: ACC.

———.1994b. The Scope of and Distinction between Articles 126 and 127 of the Treaty on the European Union and the Implementation of the Subsidiarity Principle, in Jan de Groof (ed.), *Subsidiarity and Education, Aspects of Comparative Education Law. First Report of the International Education Law Association*. Leuven: Acco.

De Groof, Jan, Neave, Guy and Svec, Jurij (eds.). 1998. Governance and Democracy in Higher Education, Vol. 2. in the Council of Europe series *Legislating for higher education in Europe*. Dordrecht: Kluwer, p. 392.

De Weert, Egbert and van Vught Thyssens, Lieteke. 1998. Academic staff between threat and opportunity: changing employment and conditions of service, in Jongbloed, Ben, Maassen, Peter and Neave, Guy (eds.), *From The Eye of the Storm: Higher Education's Changing Institution*. Dordrecht: Kluwer, pp. 39–64.

Durand-Prinborgne, Claude. 1992. France. in Clark, Burton R. and Neave, Guy (eds.), *The Encyclopedia of Higher Education*. Oxford: Pergamon Press. 4 vols. Vol. 1 *National Systems of Higher Education*, p. 217.

Enders, Juergen and de Weert, Egbert. 2004. The International Attractiveness of the Academic Workplace in Europe. Frankfurt am Main: Gewerkschaft Erziehung und Wissenschaft.

Eustace, Roland. 1987. The English ideal of university governance. *Studies in Higher Education*, Vol. 12, No. 1, pp. 7–22.

Frijhoff, W. 1992. The Netherlands in Clark and Neave (eds.), Vol. 1. *National Systems of Higher Education*, pp. 491 ff.

Garcia-Garrido, José-Luis. 1992. Spain, in Clark and Neave (eds.), *National Systems of Higher Education*, pp. 663–664.

Geiger, Roger. 1986. *Private Sectors in Higher Education: Structure, Function, and Change in Eight Countries*. Ann Arbor: University of Michigan Press.

Gellert, Claudius (ed.). 1993. *Higher Education in Europe*. London: Jessica Kingsley.

Gellert, Claudius. 1997. Innovation *and Adaptation in Higher Education: the Changing Conditions of Advanced Teaching and Learning in Europe*. London: Jessica Kingsly.

Herrin, Judith. 2006. The Byzantine University: a Misnomer. in Blückert, Neave and Nybom (eds.), *The European Research University—an Historical Parenthesis?* New York: Palgrave.

Jaksic, Ivan. 1989. *Academic Rebels in Chile: the Role of Philosophy in Higher Education and Politics*. Albany, NY: State University of New York Press.

Johnstone, Bruce D. 2004. The Economics and Politics of Cost Sharing in Higher Education: Comparative Perspectives. *Economics of Education Review*, Vol. 23, No. 4, pp. 403–410.

Kwikkers, Peter et al. 2005. *Evenwicht zonder sturing: Wegen voor nieuw hoger onderwijs en wetenschap, deel 1*. Haag: Sdu-Uitgevers.

Levy, Daniel. 1986. *Private Education: Studies in Choice and Public Policy*. New York: Oxford University Press.

Martinelli, A. 1992. Italy, in Clark and Neave (eds.), Vol. 1, *National Systems of Higher Education*, pp. 355–369.

McNair, John. 1984. *Education in a Changing Spain*. Manchester: University Press.

Neave, Guy. 1998 "Quatre modèles pour l'Université," Courrier de l'UNESO, septembre[B1] 1998.

——— 2001. The European Dimension in Higher Education: an Excursion into the Modern Use of Historical Analogues, in Huisman, Maassen and Neave (eds.), *Higher Education and the Nation State*. Oxford: Elsevier Pergamon, pp. 13–73.

——— 2003. On the Return from Babylon: a Long Voyage around History, Ideology and Systems Change, in Jon File and Leo Goedegebuure (eds.), *Real-Time Systems: Reflections on Higher Education in the Czech Republic, Hungary, Poland and Slovenia*. Brno, Czech Republic: Vuilin Publishers, pp.15–37.

——— 2004a. Higher Education Policy as Orthodoxy: Being one Tale of Doxological Drift, Political Intention and Changing Circumstances, in David Dill, Ben Jongbloed, Alberto Amaral and Pedro Teixeira (eds.), *Markets in Higher Education: Rhetoric or Reality?* Dordrecht: Kluwer Academic Publishers.

——— 2005a. The Supermarketed University: Reform, Vision and Ambiguity in British Higher Education. *Perspectives: policy and practice in higher education*, Vol. 9, No. 1. March 2005, pp. 17–22.

——— 2005b. On Prophets and Metaphors: Devices for Coping in Times of Change, in Jürgen Enders et al. (eds.), *European Higher Education and the Research Landscape 2020: Scenarios and Strategic Debates*. Enschede: CHEPS.

——— 2006. The Evaluative State and Bologna: Old Wine in New Bottles or simply the Ancient Practice of "Coupage"? Higher Education Forum, vol. 3, March 2006, Hiroshima (Japan) Research Institute for Higher Education, Hiroshima University, pp. 27–46.

——— forthcoming. Patterns. in Walter Ruegg (ed.), *A History of the University in Europe*. Cambridge: University Press. Vol. 4, *Universities in the Contemporary World*. 1945–1990.

Nybom, Thorsten. 2003. The Humboldtian Legacy: Reflections on the Past, Present and Future of the European University. *Higher Education Policy*, Vol. 16, No. 2. June.

Renaut, A. 1995. *Les révolutions de l'université: Essai sur la modernisation de la culture*. Paris: Calman-Levy.

Rothblatt, Sheldon. 1997. *The Modern University and some of its Discontents: the Fate of Newman's Legacy in Britain and America*. Cambridge: University Press.

Schwartz-Hahn, Stephanie and Westerheijden, Don. 2004. Accreditation *and Evaluation in the European Higher Education Area*. Dordrecht: Kluwer Academic Publishers.

Scott, Peter. 2000. *Higher Education Re-formed*. London: Routledge.

Trow, Martin. 2003. In Praise of Weakness: Chartering, the University of the United States, and Dartmouth College. *Higher Education Policy*, Vol. 16, No. 1. March.

Van Wageningen, Ann. 2003. De Staat van de Universiteit: een rechtsvergelijkende studie over de institutionalisiering van de universiteiten in Nederland, Frankrijk en Nordrhein-Westfalen, PhD thesis. Enschede: CHEPS.

Verger, Jacques. 1986. *Histoire des Universités en France*. Toulouse: Privat.

Weisz, George. 1983. *The Emergence of Modern Universities in France, 1863–1914*. Princeton: University Press.

Williams, Gareth. 1992. Finance and the Organisational Behaviour of Higher Education Institutions, in Edgar Frackmann and Peter Maassen (eds.), *Towards Excellence in European Higher Education in the 90's*. Utrecht: Lemma.

——— 2004. The Higher Education Market in the United Kingdom, in David Dill, Ben Jongbloed, Alberto Amaral and Pedro Teixeira (eds.), *Markets in Higher Education: Rhetoric or Reality?* Dordrecht: Kluwer Academic Publishers, pp. 241–269.

Chapter Two

Legitimating the Difference: Private Higher Education Institutions in Central and Eastern Europe

Snejana Slantcheva

Introduction

Ever since their recent establishment, private higher education institutions in Central and Eastern Europe have had to justify their existence on the higher education landscape. Paradoxically, private institutions in the region appeared in a particular historical moment in response to a legitimation crisis of post-communist public higher education institutions. State colleges and universities failed to respond effectively enough to the new challenges posed by two main sets of factors: on the one hand, the transition to democratic societies and market economies after the fall of the totalitarian regimes and, on the other hand, the powerful trends of globalization that involve economic restructuring, a changing role of the state, shifting demographics, new technologies, and increased international interdependence, which have affected higher education systems throughout the world.

Recent changes in higher education have led to a growth in private institutions worldwide. The emergence of private sectors in higher education in most post-communist countries, however, is recognized as "one of the principle developments characterizing a systemic transformation of higher education in Central and Eastern Europe" (UNESCO 2003, p. 3). Beginning in the 1990s, private institutions sprang up across the region to fill gaps in the higher education landscape formed by the increased demand for higher education, the nascent market economies, and the priorities of a spawning civil society. Within several short years, the private higher education sectors in the countries of Central and Eastern Europe grew quickly, although unevenly. In Romania alone, around 250 institutions appeared between 1990 and 1993 (Bollag 1999). In Poland, 6 private institutions were already registered by the end of

1990; and by 2002 their number reached 250; private sector enrollments of 50,000 students in 1994 climbed to more than a half million in 2001, amounting to almost one-third of the Polish student body (Kwiek 2003). By academic 2000/2001, within a decade of the appearance of private institutions, private sectors across the region enrolled a significant number of students, ranging from more than 28 percent of total student enrollments in Poland and Romania and 23 percent in Estonia, to 13 percent in Hungary and 11 percent in Bulgaria, to 4.7 percent in the Czech Republic and 0.8 percent in the Slovak Republic (table I.1 in Introduction Appendix). Although growth rates across Central and Eastern Europe have slowed, new private institutions are still appearing (table I.2 in Introduction Appendix). High private student enrollments were also characteristic in many of the former Soviet republics (table I.1 in Introduction Appendix).

The rapid establishment of new private institutions within an existing legal vacuum soon invited questions concerning legitimacy. Seen worldwide, growth in private higher education has been often sudden and surprising, largely unanticipated and unregulated, within a legal vacuum and with challenged legitimacy (Levy 2004). What is more, "private roles often emerge on the margins of what is allowed, in gray areas that policy did not foresee" (Levy 2002, p. 14). And the case is no different for Central and Eastern Europe. Private institutions have appeared to offer "more," "better," and/or "different" educational alternatives than those provided by the state (Geiger 1985). But despite the fact that, unlike the existing public institutions, these new private colleges and universities are untainted by the communist past, respond to various pressing demands of a transitional society, and embrace the major postulates of higher education reform—often with little or no direct use of taxpayers' money—they still continue to dwell "on the fringes of legitimacy" (Clark in Pfeffer and Salanchik 1978, p. 196).

This chapter explores the reasons behind the challenged legitimacy of private higher education institutions in the region. I argue that, to a large extent, the persisting search for legitimacy of Central and Eastern European private institutions of higher education is reflective of the transition in existing values in post-communist societies as a whole. These societies are in general hesitant to accept private provision of higher education, question the prioritization of human resource development over scientific research, and have historically placed more reliance on the state to be the caretaker of private goods, to think for its people and to act on their behalf.

My argument builds upon a brief analysis of the concept of legitimacy in organization theory as a symbolic representation of societal evaluation of institutional goals and the means to achieve them. These three important aspects of legitimation—legitimation of goals, legitimation of means to achieve these goals, and sources of legitimacy—guide this chapter's analysis of the private institutions of higher education in the region. As a caution, the generalizations attempted in this chapter are not intended to underplay institutional idiosyncrasies or national differences; they are made in order to illuminate the specific issues at hand. In addition, private institutions here refer to those institutions that by law are registered as private or nonstate.

Legitimacy of Organizations

On the one hand, legitimacy pertains to societal evaluation of institutional goals and core values. As Pfeffer and Salancik state, "Because organizations are only components of a larger social system and depend on that system's support for their continued existence, organizational goals and activities must be legitimate or of worth to that larger social system" (1978, p. 193). On the other hand, legitimacy also extends to the societal acceptance of the "main functional patterns of operation, which are necessary to implement the [core] values" (Parsons 1956, p. 68). More specifically, Meyer and Rowan consider the formal structure of modern organizations—the procedural aspects, their emphasis on formality, offices, specialized functions, rules, records, routines—to signal rationality, irrespective of their effects on outcomes (1977). In other words, how goals are achieved—through what processes, structures, rules, routines, and so on—carries just as much weight in societal evaluation as what these goals are. Particularly concerning organizations operating in institutional environments where "they are rewarded for establishing correct structures and processes, not [solely] for the quantity and quality of their output; . . . such organizations have a special need for procedural legitimation and are especially vulnerable to attacks on the plausibility of their work arrangements and procedures" (Scott 1991, pp. 167–169).

Organizations are expected to model in procedure and practice the core values to which they are committed. Or as Dowling and Pfeffer assert, "Organizations seek to establish congruence between the social values associated with or implied by their activities and the norms of acceptable behavior in the larger social system of which they are a part. Insofar as these two value systems are congruent we can speak of organizational legitimacy. When an actual or potential disparity exists between the two value systems, there will exist a threat to organizational legitimacy" (1975, p. 122). Hence, with respect to academic organizations schools attain legitimacy in a society to the extent that their goals are connected to wider cultural values, such as quality, transparency, fairness, and open pursuit of truth, and to the extent that they conform in their structures and procedures to established "patterns of operation" specified for educational organizations (Scott 1995, p. 21).

The sources of legitimacy—or which groups or institutions have the authority to confer their approval on an organization or its practices of a given type—are yet another important aspect. In institutional environments, organizations refer to a myriad of external actors whose evaluation impacts them. In the sphere of higher education, for instance, colleges and universities relate not only to accreditation agencies and professional disciplinary associations, but also to government bodies, state agencies, peer institutions, students, parents, employers, and donors.

Legitimacy and Private Institutions of Higher Education in Central and Eastern Europe

Since 1989, private institutions have made a strong contribution to the changing landscape of higher education in Central and Eastern Europe. Roger Geiger has identified three rationales for the existence of private alternatives to state higher education: "cases in which *more* higher education was demanded than was provided by the state; cases in which groups desired *different* kinds of schools from those provided, and cases in which qualitatively *better* education was sought" (1985, p. 387). In his study of the growth of the Latin American private sector, Daniel C. Levy documents three consequent "waves" of private growth each illustrating a specific set of private roles (1986, 2002). The first wave involved the establishment of Catholic universities with a religious role (reflecting the *different* alternative). The second wave of private institutions came as a reaction to the perceived decline in quality of public higher education as a result of "massification;" these institutions assumed elite roles (or, the *better* educational alternative). Finally, the third wave of private institutions appeared as a response to the rising demand for higher education that exceeded the public supply. These institutions assumed roles with a mixture of tasks that were related to access (or the *more* alternative). "It is mostly the third wave that foreshadowed for the late twentieth and early twenty-first century the fresh growth of private higher education in Latin America—and the startling growth in much of the rest of the developing world and the post-communist world" (Levy 2002, p. 3).

From the above-described scenarios, demand-driven institutions present the bulk of the newly created private universities and colleges across Central and Eastern Europe. In the 1990s, many private institutions of higher education were established in attempts to accommodate the rapid increase in demand unleashed by the political and economic changes, thus playing a strong role in increasing the capacity of the national higher education systems. Communist systems of higher education were, as a rule elitist. The high gross student enrollment rates at the tertiary level in the Baltic countries in 1985, of 32 percent in Lithuania, 24 percent in Estonia, and 22 percent in Latvia, were unusual for the region. For most of the countries, gross enrollment rates were below 20 percent, going as low as 10 percent in Romania in 1985 (table 2.1).

Table 2.1 Gross enrollment ratio*—tertiary level—Central and Eastern Europe

Country	*1985 (%)*	*2001 (%)*
Bulgaria	18.9	40.1
Czechoslovakia	15.8	29.8–Czech Republic 30.3–Slovak Republic
Estonia	24.2	36.4
Hungary	15.4	59.3
Latvia	22.7	39.8
Lithuania	32.5	64.3
Macedonia	24	59.1
Poland	17.1	24.3
Romania	10	55.5

Note: *Gross enrollment ratio, tertiary level is the sum of all tertiary level students enrolled at the start of the school year, expressed as a percentage of the mid-year population in the five year age group after the official secondary school leaving age.

Source: World Bank 2001.

The liberalization of state policy in higher education in the late 1980s and early 1990s stimulated the growth of the sector of higher education. Demand for higher education increased several times. To respond to this escalating demand, dozens of private institutions appeared almost overnight to offer degrees in sought-after fields, such as business administration, economics, law, computer sciences and foreign languages. Many of these institutions followed clear goals and purposes. Others were structured in strong emulation to the existing public models in their fields. Institutions established in cooperation with or as branches of public institutions were also present. Internationally, such partnership arrangements are not uncommon in countries where "academic quality and prestige traditionally reside in public universities but fresh political-economic and higher education forces create a need for very different providers" (Levy 2003, p. 28). And there were also those institutions that were founded to serve as "cash-generation engines for public universities"—channeling the student fees public universities were not able to charge by themselves (Tomusk 2003, p. 218), or as "a source of supplementary income for academics in badly financed public systems" (Reisz 2003, p. 15).

There also emerged a need for *different* institutions from those provided. Examples of different institutions of higher education in the region come predominantly from the new private sectors such as liberal arts universities and colleges, distance-learning programs offered initially by private institutions, higher professional schools, institutions established to cater to various minority populations, and higher schools of religious character. These institutions are seen as part of a larger process of democratization of the higher education landscape designed to replace what Pepka Boyadjieva has termed the "one-dimensionality" of the preceding social order with "pluralism . . . as an organizing principle of social life" (Boyadjieva 2003, p. 5). Finally, there were also private institutions established that aspired to offer *better* educational opportunities in comparison to the existing ones compromised by the communist order. One such category of private schools perhaps were, as identified by Tomusk, institutions created with the specific agenda to challenge the existing institutional order founded explicitly as being "the jump off point for a new higher education for a new society" (2003, pp. 229–230).

Whatever the specific rationale behind their establishment, private institutions in the region have had to grapple with social acceptability over time. As a rule, "the early years of the existence of organizations is the period during which they are most vulnerable to the liability of newness" (Singh, Tucker and House in Scott 1995, p. 156). And this certainly holds true for the private institutions of higher education in Central and Eastern Europe. However, we must look further, beyond the novelty argument, if we are to understand why—over 15 years since the changes—private higher education continues to occupy "the fringes of legitimacy." The persisting legitimacy deficits of private institutions across the region have been well documented by researchers. As Reisz notes in 2003, private universities in Central and Eastern European countries are still "negatively perceived by government, public universities and large sectors of the population" (p. 24). The reluctance to accept the private sector, the distrust of faculty, the government, and even of the labor market have also been recorded by Galbraith (2003). And in Bulgaria, Boyadjieva observes that "public opinion remains rather skeptical toward the quality of the education they [the private institutions] offer and the scientific criteria applied. Frequently, suspicions arise that such institutions emerge for nonacademic reasons to serve certain individual or group interests" (2003, p. 6).

Explanations to the persisting legitimacy deficits of private institutions in the region can be found in the misalignment between wider societal values and the goals and means toward goals achievement of private institutions in general. On the one hand, the overriding priority of private institutions of higher education in the region seems to be the development of human resources for post-national states increasingly characterized by global economic interdependence and multicultural civil societies—a goal that is still perceived by many to be at cross-purposes with the traditional higher education prioritization of scientific research, and the formation of a national elite. And on the other hand, private institutions' strong reliance on private means toward achieving their stated goals, such as funding primarily from private sources as opposed to the state, governance not directly under the state, and functions not defined by the state—although increasingly overseen by it—has also led to mistrust.

Yet explanations can also be found in the legitimation sources that private institutions rely on. Often founded within legal vacuums, private institutions have not always enjoyed the most powerful source of legitimacy—the regulative framework, provided by the state—which primacy has been characteristic for the "legalistic" societies of Central and Eastern Europe (Cerych 1995, p. 429). As a result, social acceptability has been preconditioned by the difficulties in reconciling this major source of legitimacy with the growing power of the market, that is students, parents, employers, and the legitimating power of Western educational models. The following analysis focuses on these issues.

Legitimating the Goal of Educating Global Citizens

As a group, private institutions across Central and Eastern Europe exhibit specific common characteristics: emphasis on student-centered teaching and learning, narrow programmatic scope, market responsiveness, practical applicability, regional engagement, and experimentation with teaching styles.

An analysis of different explicitly stated missions or implicitly expressed educational pursuits of various types of private institutions across the region helps develop these characteristics in more detail. In this manner, most private institutions claim to place the student at the center thus focusing above all on teaching and learning, or the transmission of knowledge, as their core function. Different forms of pedagogical and technical innovation are often complemented by practical training in programs that promise to produce a skilled, flexible, and critically thinking labor force. Their contribution to the enhancement of student employability is expressed in their commitment to developing transferable skills in students and offering courses, including at master degree programs, in fields demanded by the market such as business, finances, banking, law, and economics. Research is conducted mainly to support classroom teaching. Very few of the private institutions train doctoral students. For example, out of 221 institutions in Poland, only 51 are entitled to offer master degree programs and only 2 to confer doctoral degrees (Kwiek 2003), whereas the 7 Bulgarian private universities graduated 3 doctoral students in academic 2002/2003 (NSI 2003). Seen from this perspective, private institutions serve as knowledge banks tuned to the specific professional environment they focus on and respond to. In addition, they have been "not only knowledge depositories and transmitters, but also sources of solutions to many practical problems," asserting themselves as agents of development (UNESCO 2003, p. 17), involving their academic communities in the local/regional problem-solving agendas.

With respect to their institutional profiles, most private institutions in Central and Eastern Europe offer a "limited scope" of programs (Levy 1987) designed predominantly in short-term study degrees, mostly professional and at the bachelor level. The variety of these institutional establishments encompasses colleges, institutes, academies, professional schools, liberal arts higher schools, and universities. Although private institutions in some Central and Eastern European countries outnumber public institutions (see table I.2 in Introduction Appendix), most of these institutions are small, with weak infrastructures. Their corporate academic culture is somewhat diluted. A large number of their faculty is part-time, usually coming from the larger, older public institutions. Many of their students (themselves representing a rather mixed group with respect to age and social status) are also part-time, distance-learners, taking specific courses or virtual classes.

The primacy of teaching, learning, and professional orientation pursued by this group of institutions speaks to a very important shift in goal prioritization from within the bundle of the traditional functions of higher education. Parsons and Platt assert that higher education fulfills simultaneously four major functions: the pursuit of pure research and academic training (as a core function), general education, professional preparation and production of technically usable knowledge, and the promotion of cultural self-understanding (1973, pp. 90–102). Whereas Parsons and Platt base their analysis on the highly functionally diversified American system of higher education, it is important to note that such a perspective of systemic differentiation has not taken hold within the Central and Eastern European model where the traditional bundling of different functions under the roof of a single institution has prevailed. The different prioritization of core functions of higher education given in the Central and Eastern European private colleges and universities comes as a reflection of the growing functional specificity of universities "within a system of science and scholarship (*Wissenschaftssystem*) that was differentiating itself with increasing rapidity" with the advent of the modern, knowledge-based society (Habermas 1993). "Different groups with different vantage points within the university perceive the various functions as having different degrees of importance. In this way the corporative consciousness becomes diluted to an intersubjectively shared awareness that while some do different things than others, as a group all of them, insofar as they are engaged in one form of science or another, fulfill not just one but rather a whole complex of functions" (Habermas 1993).

The shift of relative weight within the bundle of higher education functions exemplified by private institutions in the region takes place against deep-seated educational values, formed above all by the philosophy of the modern German university. The idea of the university, defended vigorously by the founder of the Berlin University Humboldt and Schleiermacher, and embodied in the German research university of the beginning of the nineteenth century, posited the unity of teaching and research at the center. As Humboldt would say, the relation between Teacher and Student will therefore become completely different than before. The former is not there for the latter; both are there for science (1810, p. 256)[1] thus emphasizing the primacy of scientific research and the collaborative pursuit of truth. University studies are conducted in and through the process of research. After the political changes in Central and Eastern Europe, this ideal has been upheld more than before. "The prospect of resurrecting the Humboldtian university was one of the more appealing and much-sought-after opportunities in the fall of the Berlin wall held out to an academic profession in Eastern Europe which, for many years, regarded itself as existing in a state of 'inner exile'. Arguably, the 'inner strength' of the Humboldtian myth survived longer in the newly liberated systems of higher education of Eastern and Central Europe precisely because of its value as a symbol of resistance and a counterlegitimacy to a political order superimposed on the university from without" (Neave 2003, p. 137). Indeed, the aspirations toward returning to the Humboldtian roots have been supported by the traditional educational values in the region. In most Central and Eastern European countries, despite the separate research academies, research is allegedly the key emphasis in university missions and also a priority requirement in the accrediting regulations.

As a result, private institutions' clear prioritization of developing human resources for the new regionally and globally integrated economies and the knowledge societies, when compared to scientific research, has not been easily accepted across the region. As UNESCO's report notes, "While it is not inconceivable that certain higher education institutions (e.g., the newly established private institutions) might not include research in their mission statements, focusing only on teaching, such a development is not in line with the well-established tradition of European higher education" (2003, p. 28). And the report further notes that "Higher education institutions in Central and Eastern Europe repudiated such notions [of market employability] right after the end of communism because of the very tight association with, and subordination to, the economic system that had prevailed during the communist period . . . such a trend is also associated with more questionable degrees, with a weak academic background . . . The distinction between 'just in case training,' provided mostly by classical institutions, and 'just in time training,' offered by the newly established institutions, many of them of a transnational type, has also generated the mushrooming of new types of degrees which are more closely linked to the labor market while less so to the classical academic division of knowledge. The situation is leading to a certain tension in the academic recognition of qualifications and in the traditional system of academic values" (2003, p. 17). Perhaps an illustrative example of this tension can be discerned in the public statement of the Rector of the oldest public Bulgarian university, who noted: "Higher education does not prepare students so that they can work somewhere" (Biolchev 2004, p. 11).

Legitimating the Private Means in the Provision of Higher Education

Distinguishing between private and public institutions has not been an easy task and international research attests to it (Levy 1979, 1986; Altbach 1999). Defining what a private institution means also presents a challenge in the Central and Eastern European higher education landscape. Some legal distinction has been increasingly attempted in different countries, such as the Czech Republic, the Slovak Republic, Belarus, Estonia, for instance, where explicit division of institutions into state, public, and private took place. Recording the private/public ambiguity in general, Levy has singled out "three separate ideal-typical criteria" for privateness in private higher education, any of which might be used to identify a private institution: finances, governance, and function (1986, p. 16). Since none of these criteria is clear-cut, different degrees of privateness can be established.

Of these three criteria, private financing is perhaps the most obvious one associated with private higher education in the region. However, examples of state support, direct or indirect, for private institutions can be found in several countries. In Hungary, students can attend private colleges on state subsidy. Many private higher education institutions of religious character such as the Hungarian church-run colleges and universities, the Polish Catholic University in Lublin, the Slovak Catholic University of Ruzomberok are supported by the state. Romanian accredited private institutions are eligible for state subsidies. This subsidy is also given in Estonia, the Czech, and the Slovak Republics. Despite these examples, however, most private institutions throughout the region still rely on finances other than the state, collected primarily through student fees and/or international support.

Consistent with key private higher education literature internationally, private institutions are more private than public regarding their governance as well. Different creators can be identified behind them, from charismatic individuals (Dahrendorf 2000) as in the case of the International Concordia University in Tallinn (Tomusk 2003), to foundations (the foundation-run colleges of Hungary; the Central European University in Budapest is another example although this institution has been until recently recognized as a foreign institution in Hungary (Goemboes 2003, p. 423)), to business corporations such as the Skoda Auto College in the Czech Republic, to minority rights activists such as the multiethnic South East Europe University of Tetovo in Macedonia, to religious groups. Again, however, in most countries, the final approval of the state has been the prerequisite to official registration as a higher education institution. Lastly, to the extent that private institutions orient themselves predominantly toward the market, toward certain regional demands, or toward the interests of minority or religious groups, they are mostly private in function as well.

Setting definitions aside, one point is of great importance here. In the post-communist period, the meaning behind privateness has emphasized the particular other-than-the-state nuance that with regard to the three criteria has translated into: finances other than the state, governance with no state participation, and function not defined by the state. In other words, the important role, or rather absence, of the state in denoting privateness comes to the forefront for, especially in communist times, the state reigned supreme in setting higher education parameters. Thus, although not unique for the countries of Central and Eastern Europe, or to higher education per se, the controversy over private provision of higher education is particularly intense here due to the deep-seated values inherited from the recent past. The totalitarian society was a "one-dimensional society—a society built on the one ideology, one party, one form of property, one equalizing distribution policy" (Boyadjieva 2003, p. 5).

But the strong role of the state in the provision of higher education—noted as characteristic of the continental pattern of higher education in general (Clark 1983)—goes back even before communism. Although some of the predecessors of the modern Central European universities date back to the Middle Ages, such as Charles University—the first Central European university that according to Joseph Ben-David (1968, p. 192) marked the beginning of a movement that turned the universities into national institutions—opened in 1348, Krakow University in 1364, the short-lived Universities of Pecs in 1367, and Obuda in 1435, Academia Istropolitana in 1467, the majority of institutions in Central and Eastern Europe appeared around the nineteenth century as nation consolidating initiatives for the newly freed states after the disintegration of the Ottoman and the Austro-Hungarian Empires. Their role was to foster autonomous national cultures and to train officials for the state. As Alexander Kiossev states referring to the origins of the first Bulgarian university (which appeared in 1888), "Even before its [the university's] actual conception, it was regarded as a purely patriotic organization: not as an autonomous realm of truth, but rather as an institution securing national identity, supporting the nation and its practical needs . . . a higher school of this kind is necessary to produce civil servants, lawyers, and, most of all, teachers so badly needed by the state" (2001, p. 5). All throughout Europe, the origins of the modern university, appearing at the end of the eighteenth and the beginning of the nineteenth centuries, are closely associated with the reforms of von Humboldt in Prussia and Napoleon in France. "The modern university was a capital instrument in the construction and reinforcement in the nation-state . . . The state was the sole regulator of the higher education system and used traditional regulation mechanisms such as legislation (the daily life of institutions was strongly regulated by laws, ministerial decrees, circulars and regulations), funding, approval of study programmes, and in many cases the appointment of professor . . . This was in essence the idea of the modern university, based upon the concept of individual academic freedom but not of institutional autonomy" (Neave and van Vught 1994). The strong hand of the state can be seen in the few recorded examples when autonomous institutions were brought under the guiding umbrella of the state. Such is the case with Charles University that became a state institution in 1773, the Academy of Fine Arts in Prague that existed from 1799 and was nationalized in 1896, and, a little later, of the Free Bulgarian University (the only private secular higher education institution in Bulgaria before 1990) that existed from 1920 and was nationalized in 1939.

Higher education in the region expanded, if rather moderately, between the two world wars. Rise in the demand for law graduates to man the newly established civil services appeared in those countries that had gained independence or had been granted new territories—the Baltic countries, Czechoslovakia, Poland, Romania, and Yugoslavia (Ben-David 1968, p. 192). However, higher education was still a rather scarce resource. By 1938, Poland had 32 institutions educating close to 50,000 students, Romania had 16 higher education establishments that educated 26,000 students, and Bulgaria had 6 institutions with 10,000 students; there were 16 institutions in Hungary and 9 institutions in Czechoslovakia (Clark and Neave 1992, Slantcheva 2002). It was after World War II when the greatest expansion of higher education in the region occurred, predetermined by the need to replace the professional manpower decimated by the war, the communist drive for economic development, and the desire of party officials to train a politically reliable elite. In Romania alone, 38 new institutions were formed by 1950. New specialized institutions of technology and education were the predominant type and binary systems of higher education, differentiating between "theoretically based" universities and "vocationally oriented" polytechnics, were established in several countries.

During the communist period, higher education institutions multiplied as a result of state policy in an isomorphic way: most countries in the region had their alma mater, which carried the national identity throughout the years. Institutions appeared either as detached faculties from this major, oldest university or as emulations of it. Perhaps a major exception to this rule comes from Poland where the strong role of Catholicism could not be so easily undermined, allowing the Polish Catholic University of Ljublin to maintain its unique character although still under the umbrella of the state. As the institution asserts, "In the Communist era the University played a special role, being the only independent university in the Soviet Bloc" (history of the university at http://www.kul.lublin.pl/uk/history/).

In sum, the supreme role of the state in the provision of higher education has deep roots and institutions that associate their functioning with means other than the state face social mistrust. A different element comes from the specific connotation that the notion of privateness carried over in societies where all aspects of privateness were shunned. Most of all, it has been strongly linked to the financial side. As Galbraith notes, "there is also a lingering unwillingness to acknowledge that higher education can be other than free (an axiom now manifestly violated by the state)" (2003, p. 551). Even within the state higher education sector, elements of privateness have faced strong resistance (or for that matter legal support) as well. As Bollag has registered, "The issue of study fees, for distance or full-time studies, has been a major source of political dispute in the former Communist countries" (2002). Paid programs on top of the state-subsidized quota (still maintained in a range of countries), endorsed by the beleaguered public institutions, face social disapproval. Offering paid courses for certificates or opening institutional branches in regions of demand signal the commercialization of the sector. As one illustration, after continuous social criticism, the Bulgarian system of higher education eliminated programs for pay at public institutions in 1999 (these were reintroduced later for master and doctoral programs) and introduced across-the-board tuition fees—the latter innovation too encountered strong public opposition.

Legitimating Alternative Sources of Legitimacy

Institutions of higher education are not passive in their relationship with the environment; instead, "they may be expected to exercise 'strategic choice' in relating to their environments, or elements within them" and adjust to the changing social values, design their activities in accordance with or in challenge to projected expectations, forcefully articulate and defend their pursuits from within (Scott 1991, p. 170). The different sources of external influence on institutions of higher education can be clustered into the following groups:

- state authorities (accrediting agencies, government, law enforcement);
- professional and business groups (unions and provincial boards, religious institutions);

- students (parents, counselors);
- peer institutions;
- donors and foundations;
- international organizations and associations;
- the cultural framework (potential students, secondary school counselors, alumni, parents, media, campus neighbors, community).

Who has the right and ability to confer legitimacy on academic institutions has a special bearing on the Central and Eastern European private sectors. As Cerych observed, in Central and Eastern European societies, "In a sense, only what exists through or is backed by law can exist at all" (1995, p. 429). Although this statement may need further qualification, in at least one respect it is convincing: the legal framework, upheld by the state and its collective actors, has played the role of the main legitimating source across the region. Legal grounds permitting nongovernmental provision of higher education were given in some countries right after the fall of the regimes (as early as 1990 in Bulgaria, Hungary, Romania, Poland). However, beyond permission, a legal vacuum in terms of quality requirements, structural arrangements or operational details accompanied the first several years of functioning of many private sectors (till 1993 in Hungary and Romania, 1995 in Bulgaria, 1998 in the Czech Republic, 2000 in Lithuania and Macedonia, 2001 in Poland, and 2002 in the Slovak Republic). Upon the enforcement of different requirements, examples of private institutions closed down by governments, as in Bulgaria and Romania, testify to institutional demise due to loss of legality.

The initial existence of many private institutions outside the framework of national legislation, and thus outside the norms of legitimate educational practice, did not win them many supporters. The perception that no obvious (state) organ was monitoring their operations and activities, and above all their quality, directly attested to low academic status. The governmental and cultural reaction, although belated, was the establishment of accreditation systems all across the region. As Giesecke notes, "The irony of the quality control process in East Central Europe is that it is focused on these new innovative institutions" (1998, p. 74).

Yet another traditionally strong source of legitimacy, peer recognition, has not been available to private institutions either. In different fields, the new private institutions have faced strong resistance from traditional institutions and their professors. "Organizational legitimacy appears to be especially problematic when organizations of different distinguishable types compete for the same resources or the same activities, for the same domain . . . Public organizations may find it easier to claim social worth than their private counterparts . . . An organization which can convince relevant publics that its competitors are not legitimate can eliminate some competition. The issue of education versus commercialism is not merely of theoretical interest" (Pfeffer and Salanchik 1978, p. 201). This resistance has been overwhelmingly felt throughout many existing accreditation and legal requirements composed with the strong participation of representatives of traditional institutions. As UNESCO's report notes, "a kind of transfer of the image of public institutions onto private institutions is occurring" (2003, p. 438). From an international perspective, Levy also records the existence of nonvoluntary processes that force private institutions to emulate public institutions: "Even distinctive roles are possible but only if they gain approval of government or perhaps public bodies dominated by public university interests and norms" (2002, p. 12).

As a result, the opinion that private institutions are "less accountable to the democratic government than public institutions" (Reisz 2003, p. 30) has spread. Indeed, private institutions have looked for alternative sources of legitimacy. One such source has notably been that of international, especially American and West European, models of organization, association, and higher education. A large percentage of private institutions appeared in cooperative partnerships with such organizations, sought links to them, adopted international names, offered joint degree programs, and provided foreign modes of education. Integration into the future European Union higher education and common market has also informed the operations of many of these schools.

The market has been used as yet another legitimating source—a strategy further supported by influential international organizations in the region, above all the World Bank and its strong promotion of neoliberal policies. As Altbach noted, "The discipline of the market shapes private universities" (Altbach 2006, p. 115). Catering to students, businesses, donors and accounting to them has been used to justify institutional existence before the wider society. In aspiring to the market, institutions have even emulated it in organizational structures and government by employing administration, business strategies, financial plans, offering services. In many ways, institutional entrepreneurship—institutions based upon relevance, ruled by concepts of management, planning, control and success—has been strongly emphasized.

Concluding Remarks

"Certainly, the nation state is no longer the sole framework within which higher education in Europe operates. And since the nation state has been the crucible of the modern university, *pace* von Humboldt, this is no minor metamorphosis" (Neave 2003, p. 139). In today's global village based on knowledge, advanced educational studies and lifelong learning are becoming a prerequisite for social and individual progress. Systems of higher education will be able to successfully integrate themselves in the knowledge society only through the differentiated provision of higher education. For diversity is "the gateway to competitiveness and democracy" (Boyadjieva 2003, p. 5). In this respect, the private sectors of higher education throughout Central and Eastern Europe are one example of difference that contributes to social integration.

However, there are serious challenges that accompany the differences private institutions of higher education seek to defend and develop. The process of social acceptance of private institutions in Central and Eastern Europe will depend on their ability to confront the challenges bound up with their goals and their reliance on and responsiveness to private interests and market demands. Danger lurks in the mere replacement of a former class of state technocrats with global workforce functionaries. Combining the search for truth and knowledge creation with training of global citizens is one great challenge that these institutions will need to face. In their clear prioritization of human resource development, private colleges and universities must remain sites for the pursuit of truth, something that cannot safely be reduced simply, at the end of the day, to job placement but instead must also promote the discovery of new scientific answers to the pressing problems of contemporary society. A further challenge lies in the dependence on private interests and market demands, which tend to be self-serving. Private provision does not exclude appealing to social goals, and confronting this major challenge will be yet another hurdle on the road to recognition.

Note

1. Author's translation. The original German text reads "Das Verhältniss zwischen Lehrer und Schüler wird daher durchaus ein anderes als vorher. Der erstere ist nicht für die letzteren, Beide sind für die Wissenschaft da . . ."

Bibliography

Altbach, Ph. (ed.). 1999. *Private Prometheus: Private Higher Education and Development in the 21st Century*. Westport, CT: Greenwood Press Group.

——— 2006. Private Higher Education: Themes and Variations in Comparative Perspective (Winter 1998). In *International Higher Education: Reflections on Policy and Practice*. Boston: Boston College Center for International Higher Education, pp. 113–118.

Ben-David, J. 1968. In D.L. Sills (ed.), International Encyclopedia of the Social Sciences, v. 15. New York: McMillan Company and the Free Press.

Biolchev, B. 2004. We are Creating Clerks. *Trud*, May 25 (in Bulgarian language).

Bollag, B. 1999. Private Colleges Reshape Higher Education in Eastern Europe and Former Soviet States. *The Chronicle of Higher Education*, June 11.

——— 2002. Slovakian Universities Denounce Law that Eliminates Fees for Distance Learning. *The Chronicle of Higher Education*, April 11.

Boyadjieva, P. 2003. University Models and Social Change: American University Model as a Factor for Democratization of the Bulgarian Society. Seventh Joint Conference of North American and Bulgarian Scholars, The Ohio State University, Columbus, OH. October.

Cerych, L. 1995. Educational Reforms in Central and Eastern Europe. *European Journal of Education*, Vol. 30, No. 4, pp. 423–436.

Clark, B. 1983. The *Higher Education System. Academic Organization in Cross-National Perspective*. Berkeley and Los Angeles, CA: University of California Press.

Clark, B. and Neave, G. 1992. *The Encyclopedia of Higher Education*. London: Pergamon Press.

Dahrendorf, R. 2000. *Universities After Communism*. Hamburg: Koerber Stiftung.

Dowling, J. and Pfeffer, J. 1975. Organizational Legitimacy: Social Values and Organizational Behavior. *Pacific Sociological Review* Vol. 18, No. 1, 122–136.

Galbraith, K. 2003. Towards Quality Private Higher Education in Central and Eastern Europe. *Higher Education in Europe*, Vol. 28, No. 4, December, pp. 539–559.

Giesecke, H. 1998. In Galbraith, K., Towards Quality Private Higher Education in Central and Eastern Europe. *Higher Education in Europe*, Vol. 28, No. 4, December, pp. 539–559.

Geiger, R. 1985. The Private Alternative in Higher Education. *European Journal of Education*, Vol. 20, No. 4, pp. 385–398.

Gömbös, E. 2003. Private Colleges in Hungary—Seeking Students and Striving for Recognition: The Example of the International Business School of Budapest. *Higher Education in Europe*, Vol. 28, No. 4, December, pp. 421–431.

Habermas, J. 1993. *The Idea of the University: Learning Processes. In The New Conservatism, Cultural Criticism and the Historians' Debate*. Cambridge, MA: The MIT Press.

Humboldt, W. 1810. Über die innere und äussere Organisation der Höhren wissenschaftlichen Anstalten in Berlin (On the Inner and Outer Organization of the Higher Scientific Institutes in Berlin). In W. von Humboldt, *Werke in fünf Bänden (Works in Five Volumes)*, Editors Von A. Flitner und K. Giel, Berlin, Bd. 4, 1964, pp. 255–266.

Kiossev, A. 2001-11-02. The University Between Facts and Norms. *Critique and Humanism*, at http://www.euroscience.org/WGROUPS/YSC/BISCHENBERG prep-session4a.pdf#search=%22Alexander%20Kiossev%22, last accessed October 4, 2006.

Kwiek, M. 2003. PROPHE country data and context at http://www.albany.edu/dept/eapsprophe/data/National_Data/PROPHEGenericDataTablePoland.xls and http://www.albany.edu/dept/eaps/prophe/data/National_Data/PolandDataBlurb.doc, last accessed October 4, 2006.

Levy, D.C. 1979. *The Private-Public Question in Higher Education: Distinction or Extinction? Higher Education Research Group*. New Haven, CT: Yale University Press.

Levy, D.C. 1986. *Higher Education and the State in Latin America: Private Challenges to Public Dominance*. Chicago: The University of Chicago Press.

——— 1987. A Comparison of Private and Public Educational Organizations. In Powell, W. (ed.), *The Nonprofit Sector, A Research Handbook*. New Haven and London: Yale University Press, pp. 258–277.

——— 2002. Unanticipated Development: Perspectives on Private Higher Education's Emerging Roles. PROPHE Working Paper #1. PROPHE Working Paper Series at http://www.albany.edu/dept/eaps/prophe/publication/paper/PROPHEWP01_files/PROPHEWP01.doc, last accessed October 4, 2006.

———. 2003. Profits and Practicality: How South Africa Epitomizes the Global Surge in Commercial Private Higher Education. PROPHE Working Paper #2. PROPHE Working Paper Series at http://www.albany.edu/dept/eaps/prophe/publication/paper/PROPHEWP02_files/PROPHEWP02.doc, last accessed October 4, 2006.

——— 2004. The New Institutionalism: Mismatches with Private Higher Education's Global Growth. PROPHE Working Paper #3. PROPHE Working Paper Series at http://www.albany.edu/dept/eaps/prophe/publication/paper/PROPHEWP03_files/PROPHEWP03.doc, last accessed October 4, 2006.

Meyer, J.W. and Rowan, B. 1977. Institutionalized Organizations: Formal Structure as Myth and Ceremony. *American Journal of Sociology*, Vol. 83, No. 2 (September), pp. 340–363.

Neave, G. 2003. On Scholars, Hippopotami and von Humboldt: Higher Education in Europe in Transition. *Higher Education Policy*, Vol. 16, No. 2, pp. 135–141.

Neave, G. and van Vught, F. (eds.). 1994. *Government and Higher Education Relationships Across Three Continents. The Winds of Change*. London: Pergamon Press.

NSI: National Statistical Institute. 2003. *Education in the Republic of Bulgaria*. Sofia: NSI.

Parsons, T. 1956. Suggestions for a Sociological Approach to the Theory of Organizations. *Administrative Science Quarterly*, 1, pp. 63–85.

Parsons, T. and Platt, G.M. 1973. *The American University*. Cambridge, MA: Harvard University Press.

Pfeffer, J. and Salanchik, G. 1978. *The External Control of Organizations: A Resource Dependence Perspective*. New York: Harper and Row.

Reisz, R.D. 2003. Public policy for private higher education in Central and Eastern Europe. Conceptual clarifications, statistical evidence, open questions, HoF Arbeitsberichte 2/2003, Wittenberg: Institut für Hochschulforschung, Martin Luther Universität Halle Wittenberg (Institute for Higher Education Research at the Martin Luther University Halle Wittenberg).

Scott, W.R. 1995. *Institutions and Organizations*. Thousand Oaks, CA: Sage.

——— 1991. Unpacking Institutional Arguments. In W. Powell, and P. DiMaggio (eds.), *The New Institutionalism in Organizational Analysis*. Chicago: The University of Chicago Press, pp. 164–183.

Slantcheva, S. 2002. The Bulgarian Academic Profession in Transition. In Ph. Altbach (ed.), *The Decline of the Guru: The Academic Profession in Developing and Middle-Income Countries*. Boston: Boston College Center for International Higher Education, pp. 281–305.

Tomusk, V. 2003. The War of Institutions, Episode I: the Rise, and the Rise of Private Higher Education in Eastern Europe. *Higher Education Policy*, No. 16, pp. 1–26.

UNESCO. 2003. Report on Trends and Developments in Higher Education in Europe. UNESCO, at http://portal.unesco.org/education/en/file_download.php/70d8113ae74dd89803d980abd74a1abcEurope_repE.pdf, last accessed October 4, 2006.

World Bank. 2001. Educational statistics at http://devdata.worldbank.org/edstats, accessed February, 2005.

Chapter Three
Legitimacy Sources and Private Growth in the Post-Communist Context

Marie Pachuashvili

Introduction

The establishment and growth of private sectors of higher education reflects a major development within the higher education systems of many post-communist states. Development patterns, however, have been highly diverse across the region. In general, the differences in the trajectory of higher education restructuring are profound. The decline of strict central control over higher education has yielded markedly different forms of governmental steering of higher education. In fact, the differences are largely a reflection of the particular balance of the authority that the state has nevertheless retained and the influence of the growing number of stakeholders. In the same way, if the communist state was the only actor authorized to confer legitimacy on institutions, today the number of relevant actors has increased, sources of legitimacy have diversified, and vary from country to country. Most generally, societal support for the newly emerged private sectors still remains marginal. The state and its agencies continue to play a leading role in legitimizing institutions of higher education. But evaluations by peer organizations and professional associations have been gaining more weight. The main argument of the chapter is that differences in the balance amongst the sources of legitimacy are largely responsible for producing diverse growth and development patterns of private higher education as well as multiple types of legitimacy.

In most countries in the region, private sectors evolved from virtually nonexistent to substantial development. Observed variations in the sources of legitimacy on the one hand and different growth of private institutions on the other hand give a rare opportunity to explore the relationship between private growth and organizational legitimacy. In addition, the countries in the region share much in communist legacy as well as political-economic factors that have opened the possibility for privately provided education. All these provide fertile grounds for the development of new hypotheses and testing those predicted by both private higher education and organizational sociology literature (DiMaggio and Powell 1983; Levy 2004). For example, new institutionalism literature predicts that following the collapse of communism, post-communist countries would search for ideas and developmental models in neighboring nations (DiMaggio and Powell 1983). Yet, empirical evidence suggests that, having been influenced by national forces, higher education reforms became highly differentiated soon after the changes of 1989. Governmental policies toward private provision of higher education are a clear indication of such differences. It should be noted that many factors, both at the political-economic and the higher education levels were comparable across the region. These included generally low pre-transition higher education enrollment levels and the inability of public institutions to meet unleashed student demand for higher education sometimes coupled with nonrestrictive regulatory environment ensuing from initial political-economic disturbances and a loss of legitimacy of the state itself. Private institutions thrived in settings where most of these conditions converged. Providing expanded access to higher education has been probably the major legitimizing factor of private higher education worldwide in recent decades and it also seems to be central to private development in the former communist states. At the same time, the post-communist experience highlights the diversity of the sources that contribute to legitimacy as well as the different types of legitimacy that these sources lead to.

The scope of this chapter extends to countries that have emerged from communist rule in Central Eastern Europe and the former Soviet Union. The section to follow discusses the concept of organizational legitimacy, the three bases of legitimacy identified by Richard Scott, and the sources behind them. Exploring the link between the diverse sources of legitimacy on the one hand and patterns of private higher education growth and organizational goals on the other hand forms the subject matter of the following two sections. Next, drawing on empirical evidence, the section after explores some of the strategies that private institutions adopt in their pursuit of legitimacy when the legitimacy of the state and its institutions is brought into question. It considers how legitimacy is created by post-communist private institutions, what symbolic value they attempt to highlight and what strategies they use in pursuit of those aspirations. The key findings are summarized in the concluding section.

The Notion of Legitimacy

Organizational legitimacy is perceived not merely as an additional resource necessary for organizational survival and growth but also as a symbolic value created by institutions and made visible to those in a position to evaluate it. Legitimacy, according to Richard Scott, "is not a commodity to be possessed or exchanged but a condition reflecting perceived consonance with relevant rules and laws, normative support, or alignment with cultural-cognitive frameworks" (Scott 2001, p. 61). Scott distinguishes three bases of legitimacy: Regulative, Normative, and Cultural-Cognitive. The regulative perspective refers to an organization's compliance with relevant legislation, the normative element emphasizes the organization's conformity with normative evaluations by external actors, while the cultural-cognitive aspect stresses cultural support.

These three bases of legitimacy are rooted in different institutional sources of legitimacy that vary depending on context, time, and the type of organization under assessment. Thus, following Scott's classification, an institution gains legitimacy if it satisfies at least one of the following conditions: is established and run in agreement with relevant laws, licensing and accreditation requirements, enjoys normative support from students and their parents, different intermediary agencies, peer organizations, and other stakeholders, and has cultural support from the society at large. Moreover, not only do diverse bases of legitimacy shape the overall legitimacy of organization but also different combinations of legitimation sources lead to multiple types and mixes of legitimacy.

Sources of Legitimacy and Post-Communist Private Higher Education Growth Patterns

As noted, higher education in communism was overwhelmingly state controlled. Although post-communist states continue to play an important role in higher education, they are no longer the sole authority. The group of stakeholders that have gained legitimate influence over higher education has included students and their parents, international and donor organizations, peer institutions and academic oligarchy, religious groups, and business community, intermediary bodies, political parties, and other interest groups. Higher education systems in the post-communist countries thus have been moving from central control to control through intermediary bodies, the mode of which depends on the particular balance between the state and internal and external interests in each country. Correspondingly, the sources through which institutions build their legitimacy have diversified, even if conformity with state regulations continues to be important. This is especially true for private institutions that rarely rely on a single source of legitimacy. Among the actors who evaluate academic institutions, professional organizations and intermediary bodies are the most prominent along with the labor market that constitutes a potent source of legitimacy for private institutions. However, particular mixes of various sources and types of legitimacy vary significantly from country to country. Disproportional growth of private higher education across the region is largely a reflection of those differences.

Different factors that were at work in the wake of the regime change contributed to the rapid growth of private institutions even if they lacked an ample measure of legitimacy. Sociologists claim that organizations need legitimacy for their continued existence and growth. However, post-communist evidence indicates that in situations in which the public institutions remain selective and their enrollments low, strong legitimacy is not needed for private institutional growth. What is more, too rapid and easy expansion can itself undermine organizational legitimacy.

In Central and Eastern Europe, the most intensive expansion of private sectors during the first decade after the political changes took place in Romania, Poland, and Estonia (tables I.1 and I.2 in Introduction Appendix).[1] From the former Soviet republics, comparable private growth has been observed in Georgia. In all four countries, private institutions had neither an adequate measure of regulative legitimacy, since much of the growth took place in an unregulated environment immediately after the collapse of communism, nor societal support, as three out of the four countries had no tradition of private institutions.[2] In spite of this, the private sectors in these countries grew fast to accommodate almost one-third of all student enrollments soon after their appearance.

In Romania, Poland, Estonia, and Georgia, however, the fast private sector growth has been rooted in somewhat different patterns of legitimacy granted by different legitimation sources. Clearly, expanding access to higher education was the key to the private organization growth in Romania—a country with Europe's lowest (7 percent) pre-transition higher education enrollment rates. Already by 1993, Romania's total student enrollment had doubled. The role played by the newly-emerged private institutions in this enrollment growth was significant. Thus Romanian private institutions expanded with support from the market, in the face of students and their families, and to some extent, the society at large. The same was mostly true for Poland, which even with its much higher pre-transition student participation level of 16 percent still remained an elite system. But Polish private higher education institutions also enjoyed a more coherent regulative basis. As with privatization politics in general, the stance taken by the Polish government toward higher education privatization was distinctly liberal. It found an expression already in the 1990 Higher Education Act and provided legislative base for private higher education institutions. It was the lack of any regulation rather than coherent liberal policies conceived at the governmental level that gave the opportunity to Georgian private institutions to thrive. Until recently, licensing procedures have remained ill-defined while quality control mechanisms have been absent altogether. Thus, like in Romania, the market served as the major legitimizing factor for Georgian private institutions. In addition, the unparalleled economic and political downfall severely affected public institutions of higher education, which experienced a sharp fall in academic standards and faced equally serious legitimacy challenges. The Georgian example illustrates how external circumstances—in which the public sector or the state itself lose legitimacy—lower the threshold of legitimacy thus offering a chance for private institutions to grow (Levy 2004). By contrast, Estonian pre-transition student enrollments of 36 percent had been amongst the highest in the entire region. The key factor behind Estonian private higher education growth lies in the large Russian population in this country. This large Russian minority was left out of academia after the language law establishing the Estonian language in higher education was passed.[3] Accordingly, Estonia's private institutions gained initial legitimacy more by satisfying ethnic group demands or providing alternative to communist education, rather than by merely absorbing the demand that is unmet by government provision.

Regardless of the differences in legitimating sources, unrestricted private expansion provoked re-active governmental regulatory measures in all four countries. For example, Romania introduced a national accreditation procedure in 1993 that resulted in relative decline in the number of private institutions. The Accreditation Law was shortly followed by the Education Law that tightened regulation to the extent that the Romanian parliament's approval became a prerequisite for the establishment of new private institutions. Since then, strict requirements for granting authorization have been set (Reisz 2001). Georgia, the country that distinguished itself with its laissez-faire attitude to private proliferation, has been adopting more stringent policies since the government change in 2003. As a result of the 2004 revision of licensed higher education institutions, only 117 out of 227 satisfied nominal requirements for material base, space, and staff. Majority of the institutions that were not allowed to admit students for academic 2005/2006 were private. Thus, both in Georgia and Romania, delayed regulatory measures resulted in the significant decrease in institutions that had had an easy start. These measures also contributed to an enhanced social standing of surviving institutions (Levy 2002).

Some post-communist nations, by contrast, had stricter criteria for private establishments from the outset. In the settings where private institutions were permitted but strictly regulated, private institutions grew slower. There is no evidence that these private institutions enjoyed higher acceptance from the respective societies but compliance with stringent licensing and accreditation procedures, mostly tailored to public institution standards, gave them an advantage in asserting their right to exist. In countries throughout Central Asia and in Belarus, private institutions have mostly relied on regulatory legitimacy. By contrast, in Hungary, Latvia, and Lithuania, legitimating sources have been more wide-ranging and have included various interest groups, professional associations, and newly established intermediary bodies. The assessments by external committees and interest associations, mostly composed of public institutional elite, has been playing a more prominent role in higher education governance in general. In Lithuania, for example, institutional interests directed at preserving public monopoly have been identified as one of the main reasons for not allowing private institutions to operate up until 1998 (Mockiene 2001). Revealingly, it was not unleashed student demand that played a decisive role in swaying governmental policies toward allowing other than publicly provided education but the interests of various groups on the supply-side. Vilnius Saint Joseph Seminary—Lithuania's first private institution—was founded at the proposal of Vilnius Archdiocese in 1999. Out of four private universities that existed in Lithuania by 2001, three were established by religious organizations and only one—the International School of Management—with foreign involvement (Higher Education in Lithuania 2001). Requirements for establishing a new institution or a new program almost do not differentiate between public and private universities and are set so high that few providers—especially those without financial assistance from international, religious, or other donors—can meet them. The legitimacy of the private institutions in turn becomes rooted in religious or international links and thus is quite different from that of the "demand-absorbers." In general, Lithuania's highly controlled private organizations are so competitive that their tuition fees rank much higher (sometimes three and more times higher) than those for the equivalent programs in public universities.[4] Regulations, thus, can be seen both as inhibiting private growth and as conferring legitimacy to those who meet the requirements. Whether delayed or not, regulative measures such as accreditation and licensing carried out by ministries and intermediary agencies constitute one of the main legitimizing sources of post-communist private higher education.

State Control and Post-Communist Private Higher Education Goals

The type and extent of state regulation influences not only private sector growth but institutional goals as well. The role of the state and its agencies in legitimizing, and often prescribing private institutions' goals—be that of demand-absorbing or satisfying differentiated demand for religious, ethnic, and other types of education—remains considerable throughout Central and Eastern Europe and the former Soviet republics. Research has identified three major types of private institutions' goals responding respectively to demand for *better* (elite types), *different* (mostly Catholic Christian), and *more* (demand-absorbing) education (Levy 1986, 1989; Geiger 1986). As Levy (2002) recognizes, the distinguishing characteristic of the post-communist private development is the simultaneity of all three motives behind private growth. Although the balance varies, three "ideal" types of institutions serving different purposes can be identified in most of the Central and Eastern European countries.

Expanded Access

Expanded access has been a major rationale behind post-communist governments' willingness to facilitate private educational organizations. That the majority of private institutions, regardless of their mission and type, are privately funded highlights the need of governments to access nonstate recourses.[5] But significant differences exist in different government's willingness to delegate this role to the private sector.[6] For instance, Hungary had Europe's second lowest pre-1989 higher education enrollment levels. The state took it as its own responsibility to expand participation in higher education. Hungary's total enrollments more than doubled as a result of the massification process that took place in the public sector while private growth was contained through legal restrictions. Hungarian private institutions thus had to pursue different routes to legitimacy such as providing different-from-the-state education or catering to various religious groups.

Demand-absorbing institutions, by contrast, fulfill the goal of offering expanded access and/or consequent increased opportunities on the labor market. A large share of small *demand-absorbing* organizations has been characteristic mostly to those communist successor countries that have allowed unrestricted private expansion. As with similar demand-adsorbing private institutions internationally, organizations from this group generally are small nonuniversity, tuition-dependent establishments that are vocationally oriented or concentrate on some other low-cost, high-demand programs (Levy 2002). But in spite of their pervasiveness, demand-absorbing institutions do not enjoy much societal support. Their sole focusing on developing skills and credentials, rarely combined with conducting research and transmitting scientific knowledge, has been rarely perceived as higher education in the traditional sense. They are usually subject to most severe criticism from professional groups and government authorities. However, the easy formation and growth of private establishments in some countries cannot be taken as an indicator for their being less illegitimate. Quite the opposite: demand-absorbing institutions throughout the region have faced some of the most compelling legitimacy challenges; but some post-communist states, welcoming private influx into resource-starved higher education sectors, have permitted their expansion.

In more controlled settings, private institutions had to build their legitimacy through concentrating on other than demand-absorbing goals. These include elite, ethnic, and religious factors, although, as the examples below show, the set of missions that private institutions have been free to pursue has often been constrained by the state.

Ethnic and Religious Dynamics

The most illuminating example of how national forces shape organizational missions is offered by the Baltic States. In numerous respects, Latvia, Lithuania, and Estonia were more comparable cases from among all Soviet republics. These countries' exit from communism and their consequent political-economic transformation paths also bear significant parallels. But considerable differences in general politics of the three states became pronounced in 1992–93. The ethnic composition of each of these countries was probably one of the most decisive factors in determining the route into which the country's politics would evolve (Henderson and Robinson 1997).[7] The same was true for higher education policies. Although higher education restructuring attempts initiated during the immediate aftermath of the collapse of communism were comparable across the Baltic countries, the policies became highly differentiated shortly after. That the countries sharply contrast on the patterns of private higher education development illustrates this point well. For instance, in 1999, when Lithuania's first private institution gained state recognition, the private sector accounted for 25 percent of Estonia's and 12 percent of Latvia's total enrollments (table I.1 in Introduction Appendix).

The ethnic politics have played significant part in shaping governmental policies toward private education. Traditionally hostile to Moscow, all three Baltic States adopted language laws soon after the regime change according to which higher education in public sectors could only be provided in the respective native language. Therefore, it is not unexpected that Russian language providers were the majority among the first founders of private institutions. National sentiments were equally strong in each country, but Latvia and Estonia, where native populations barely constituted the majority, permitted the establishment of private educational organizations thus allowing the private sector to serve the emerging needs of the countries' minorities that were excluded from academia (for it would have been hard to justify preventing almost half of the population who did not speak an official language from participating in higher education altogether). Lithuania, which had a far smaller Russian minority drifted toward harder nationalist strategies in general. It appears, on the one hand, that exclusivist policies played a strong part in rejecting a license to non-Lithuanian providers of higher education during the times when "democracy and nationhood were competing logics" (Stepan 1994).[8]

The religious motive, on the other hand, has been crucial to the permitting and growth of private institutions in ethnically less heterogeneous Lithuania. As already noted, out of Lithuania's four private universities that existed by 2001, three were religious.[9] Institutions that focus on distinct goals have been prevalent in those post-communist countries that have used restrictive strategies toward private growth at large but have allowed and legitimized certain goals. For instance, out of the 14 percent of the students that attended private institutions in Hungary, 5.4 percent studied in religious institutions.[10] In contrast, despite Georgia's ethno-linguistic and religious diversity, the role of religion in the private sector dynamics remains marginal, not only measured in the share of total private enrollments but also in real terms. Until recently, the private sector included one Orthodox Christian Academy founded by the Georgian Patriarchate.[11] After two unsuccessful attempts in 1991 and in 1994 by the Catholic Church, an authorization was finally granted to the Sulkhan-Saba Orbeliani Institute of Theology, Culture, and History. The institution started functioning in academic 1997/98 as the first non-Orthodox Christian institution. An example offered by Romania is equally revealing. With around 87 percent Eastern Orthodox Christians, Romania is a relatively religiously homogeneous country. Despite this, out of the 12 religious private institutions operating in Romania, none concentrates on the Orthodox Christian mission. Instead, three are Roman-Catholic, three Greek-Catholic, two Protestant and four neo-Protestant educational establishments (Reisz 2001).

Several reasons might explain this apparent contradiction. First, both in Romania and Georgia, where unsatisfied demand for higher education has driven private growth, suitable conditions have existed for demand-absorbing private institutions to survive and grow. Comparatively, in Lithuania, which had one of the highest pre-transition enrollment rates of 28 percent, and in Hungary, in which expanding access became the objective of the state, institutions have been compelled to pursue distinct goals and means to legitimacy. Furthermore, that theological studies have been widely offered at public sectors of Romania and Georgia to some extent obviates the need for their private provision. However there is a major difference between the two cases. Romania clearly has been more receptive to religious diversity, while in Georgia, the environmental niche for the private sector has been further restricted by state ideology. The excessive exclusivist-nationalistic policies, underlying Georgia's politics during the first phase of the transformation, have made the existence of other than Orthodox Christian institutions hardly possible. In depth studies of both Georgia and Lithuania further suggest that the state constrains to ethnic-religious organizational goals have been in accord with values and choices of the societal majority. State authorities in each country, in fact, have often evoked prevalent and well-expressed societal discontent to justify the policies limiting pluralism in organizational goals. Belarus presents an even more visible example of the state restraining alternative goals. Viewing private initiative as a threat to society and its values, the Belarusian government openly denounces the existence of institutions concentrating on something else than what the public sector has to offer. It is revealing that with the advent of Lukashenko's authoritarian regime, almost half of the established private universities were closed down, the most recent being the European Humanities Institute.

Elite Dynamics

While elite motive accounts for the smallest share of the sectoral growth, societal prestige of the so called *elite private institutions* has been greater across the region. The factors contributing to the perception that elite institutions offer better quality of education include: provision of post-graduate programs, affiliation with various international agencies that are usually in the role of cofounders, high employability rate of graduates, and high labor market evaluations. However, most of the post-communist elite private institutions serve the same pragmatic mission and concentrate on teaching (often in selected fields) rather than on research, like their demand-absorbing counterparts. In light of this, these comparatively elite institutions have been sometimes referred to as "*semi-elite*" (Personal communication with Daniel C. Levy, April 11, 2005).[12]

Multiple Legitimacies: Strategies Employed by Post-Communist Private Institutions in their Pursuit of Legitimacy

Top private institutions of post-communist countries employ different means to legitimacy in their search for greater social standing.

Challenging Regulatory Legitimacy

As the section on the growth patterns has demonstrated, the nature of governmental policies toward private institutions was largely determined during the first phase of the political-economic transformation. In countries with major private sectors such as Romania, Georgia, Estonia, and Poland, much of the growth took place during the first half of the 1990s. The initial expansion in most cases was followed by belated regulatory action that had only partial success (Levy 2002).[13] The beginning of the 1990s were formative for democratic institutional building when most crucial choices of economic and institutional design were made. It also was the time when the authority of the state was most discredited. Thus, post-communist private expansion often took place against the backdrop of largely discredited state authority. It is not unexpected that, in their pursuit for legitimacy, the first private institutions emphasized their thrust to challenge the existing order.

In Estonia, for instance, the beginning of 1990s witnessed a spectacular growth of private institutions. Many of them carried out an ambitious and distinctive mission to challenge and substitute for the existing institutional order (Tomusk 2003). Established in 1988, the Estonian Institute of Humanities was among the first educational institutions to assert the values of institutional and academic autonomy. In Lithuania, with the active involvement of the Lithuanian Diasporas, the Vytautas Magnus University was reestablished in 1989. Although a public university, it also emphasized its independence from state control and ideology and its commitment to the symbolic values of academic freedom. It is interesting to note here that one of Lithuania's oldest educational establishments served the same purpose (in a country that did not allow private institutions) that the newly founded private university in Estonia did.

But there are important exceptions to this pattern. First, the state did not lose its authority in all former communist countries to the same extent. While we cannot discuss in detail the variations in communist regime and the countries' exit modes from it, we should note that those factors have largely determined the degree to which the ruling communist elite and the state itself retained legitimacy. Hungary is a good example of a country where the attempts of the reformist faction of the communist party ensured that some faith in the state was sustained. The fact that the party-state did not collapse and institutions were not destroyed but rather legally transformed can explain much of state-university relations (Stark and Bruszt 1998). It explains why the legitimacy of the state in fulfilling the mission of human capital development was never brought into question by the policy elite even after the regime changes, the way it happened in a number of post-communist countries. It also explains why isomorphism with public institutions rather than diversity from them has been the preferred route to legitimacy for Hungary's private institutions (Levy 1999).

But even more important to the theme of the chapter is the second point that follows from observing the evolution of organizational goals in the post-communist setting since the beginning of the 1990s. As it appears, not only do different institutions seek different legitimacies but also the same institutions aim at different legitimacies depending on the time and the circumstances. The Estonian Institute of Humanities is a revealing example of the shift in the symbolic value that institutions emphasize at different times. In its new aspiration to academic legitimacy, Estonia's first private university merged with the (public) University of Tallinn in 2005. Thus, if in the early years of the transformation the university tried to build its legitimacy by emphasizing its diversity, later on, isomorphic behavior, or even integration with a public institution, became the favored means toward academic legitimacy.

Challenging Normative Legitimacy

In the post-communist context, examples of private institutions building their reputation not by coping but against traditional and, more often than not, prestigious institutions are multiple. In an attempt to differentiate themselves from state universities, for instance, Georgia's top private institutions have set a certain level of admissions and academic standards. Such a strategy is unthinkable for paid education at public institutions. It is true that candidates applying for self-financed student admissions at public institutions should pass one entrance examination, but in reality those who can meet the cost of the study get accepted (Gvishiani and Chapman 2002). More striking in the Georgian setting are the private sector drop-out rates. If examples of public universities refusing fee-paying students are virtually nonexistent (i.e. drop-out rate is close to zero), graduation requirement standards in some privately owned institutions are so high that only about 60 percent of those initially enrolled stay on to graduate (Sharvashidze 2002). Corruption in the public sector is an additional factor: passing grades in the newly established top private institutions cannot be obtained in exchange for bribes as is the case in many traditional public universities.[14]

As in many other spheres of public life, corruption has been endemic to the Georgian public higher education sector. According to a recent study, as much as an estimated 80 percent of the Tbilisi State University entrants pay bribes to gain admission to this most prestigious institution of Georgia (Kandelaki 2005). There is ample and compelling evidence of bribes offered not only in exchange for admittance to institutions but for passing grades throughout the study process (Gvishiani and Chapman 2002; Janashia 2004; Kandelaki 2005; Corso 2005). As evidenced by other country cases, Georgia is by no means an exception in the post-communist world where corrupt practices and extensive patronage networks are pervasive in higher education. For instance, one of the rationales behind launching national entrance tests in Latvia and Lithuania was to curb bribery and other dishonest practices associated with entrance examinations. The level of corruption and clientelism is said to be so significant in the Georgian and

Russian higher education systems that both countries are following the Baltic example in introducing national entrance tests (Altbach 2005; Kandelaki 2005; Corso 2005). Academic corruption has been documented in the Ukrainian higher education system as well. It is noteworthy that many of the Ukrainian private institutions, similar to the Georgian, have chosen to move away from clientelist and nepotistic arrangements pervasive in the traditional universities and thus foster academic culture different from the existing one (Stetar, Panych, and Cheng 2005).[15]

The above examples stand in opposition to the proposed rationalization of the significant increase in the documented academic dishonesty according to which the empirical change to some extent should be attributed to *massification* and *marketization* of higher education and their corollary shift from traditional academic to commercial values (Altbach 2005). The post-communist evidence where private institutions have been trying to build their reputation and gain legitimacy not by copying but by differentiating from traditional public institutions is equally at odds with the core premises of the new institutionalism (DiMaggio and Powell 1983; Levy 2004).[16] These contradictions, however, are easily resolved if we take into consideration the particular context; the fact that corruption to a certain degree was characteristic to all communist regimes and was especially strong in countries in which *patrimonial communism* prevailed (Kitchelt 1995; Kitchelt et al. 1999).[17] It is not surprising that the strategies that the emerging private institutions have chosen in asserting their own distinctiveness and effectively competing with well-established public universities include the challenge to the practices inherited from the communist era.

In Lithuania, academic corruption is less of a concern but here too, institutions have explored diverse paths to organizational legitimacy. The International School of Management was founded in 1999 as a for-profit university. In 2004, its legal status was changed to nonprofit (due to a mistake made in the Ministry of Education). It is notable that the university leadership has made an effort to regain formal for-profit status even though the distinction between the two forms of organization is almost inconsequential in Lithuania, as in the post-communist setting in general. Hence, the question of the legal status was only of a symbolic significance. According to the institutional representatives, by choosing to declare profits earned from educational activities, they sought to emphasize the institution's distinctiveness from their public counterparts whose leadership commonly exploits the nonprofit status for its own benefits (interviews undertaken with the university representatives, September 14, 2005). As literature has acknowledged, the reason why education is one of the major services provided by nonprofit organizations throughout the world is "contract failure"—where conditions for private for-profit organizations are lacking due to the asymmetry of information between producer and consumer (Hansmann 1987). It appears that in Lithuania, where the self-serving nature of professional organizations has become a public knowledge, it is the for-profit status that serves as a defense against information asymmetries.

International Links

The post-communist evidence strongly supports the observation that the way to legitimacy commonly taken by private institutions, especially when the state has lost legitimacy is establishing links with international organizations (Levy 2004). Examples from most countries in the region of private institutions having international organizations among their founders and donors abound. Those are mostly elite and semi-elite institutions that enjoy higher social standing.

Success in the Labor Market

Lastly, the market can serve as a powerful source of legitimacy for private higher education organizations in situations where the state and traditional public institutions have lost legitimacy or when other powerful and overriding forces are at work. In the post-communist setting, as it was noted, such overriding forces include generally low pre-1989 enrollment levels, extreme focus on technical fields and absence of instruction in social sciences during communism, and fall in quality standards in the public sectors of many formerly communist nations.

That the job placement rate for Georgia's top private institutions is much higher than that of public universities is indeed indicative of their success in the labor market. The average job placement figure for the top five private institutions is 90 percent (Sharvashidze 2002). In contrast, according to the official statistics, only 3–4 percent of Tbilisi State University graduates find jobs during the first year after graduation while the figure for the Georgian Technical University, with its highest employment rates among the public institutions, constitutes 26 percent (State Department for Statistics, 2004).[18] Given the corrupt practices pervasive in public institutions and in state agencies, which has made getting a license to operate within virtually anybody's means, it is understandable why a diploma from a top private institution constitutes the most powerful signal for a potential employer. Thus attempts to foster alternative academic culture form the basis for labor market trust and recognition and serve as another important legitimizing aspect of private institutions.[19] Romania has established more formal legitimizing procedures than Georgia such as strict authorization requirements and quality assurance mechanisms that have surely contributed to the private institution legitimacy, but, as Reisz (2001) notes, labor market evaluations and recognition constitute the strongest factors that have pushed private institutions up from the bottom of the prestige hierarchy there as well. In his words "after so many years of being in business, graduates of private universities are already on the labor market since five years and have proven to be able to stand up to their position next to those pubic institutions."

Concluding Remarks

This chapter is an exploratory effort to identify salient development patterns in post-communist private higher education and examines their correlation with organizational legitimacy. A complete account of the interconnection between the two would require in-depth examination of carefully selected cases. But several important points emerge even from this cursory look. First, the study shows how varied the sources and types of legitimacy for private institutions can be, even when countries experience similar factors both at higher education and political-economic levels. Although the state has remained a major institutional source for legitimation, institutions no longer need to rely solely on state authority but can turn to diversified sources for legitimacy. A more nuanced analysis shows that the particular mix of multiple sources and types of legitimacy is largely determined by national factors and differences that exist with respect to the communist regime types and the mode of political-economic transformation.

The market is an especially strong legitimation source for private institutions in countries where the state has adopted laissez-faire stance toward private growth. Political chaos and sharp economic decline endemic to many countries during the early 1990s provided the most suitable grounds for emergent private institutions to prosper even in the absence of an appropriate regulative base. Given the inability of post-communist governments to meet high pent-up demand on higher education, these institutions were content with the success on the job market. Institutions flourished without having an adequate measure of formal legitimacy and sometimes challenging regulative and normative bases of legitimacy were employed as an important strategy toward achieving higher standing in the academic prestige hierarchy. By contrast, the goal of demand-absorbing carried less validity in countries where private institutions had to comply with stricter regulation and meet higher academic standards from the outset. Ethnic-religious composition of a given country, general politics toward and societal acceptability of diverse goals have played a major role in determining the path that emergent institutions would pursue toward the justification of their existence.

Lastly, it is notable that sources and types of legitimacy vary not only from country to country but also from time to time. While in the early 1990s, private institutions emphasized their diversity and even challenged prevalent institutional order, the same institutions presently favor isomorphism as the route to academic legitimacy. Post-communist private higher education has a history of less than two decades but there is clearly a pronounced shift in organizational goals.

Notes

1. Table I.1 in Introduction Appendix shows that the share of private sector participation has been declining in countries with largest private sectors while in other countries it has been on the rise.
2. Poland provides an exception where some religious private institutions continued their de facto existence even during communist times.
3. By 1989, the ethnic Russians constituted 30 percent of Estonia's total population.
4. To compare, study fees are comparable in Georgia's two sectors of higher education.
5. A fundamental characteristic of post-communist private higher education institutions is their extreme reliance on private resources. It is mostly religious institutions that receive finances from respective governments and religious organization, and so called "elite" institutions that obtain other than private resources, chiefly from their international cofounders. In some countries, grants and loans are allegedly available for students attending accredited private institutions.
6. This decision in turn is influenced by forces at the national level, probably with macroeconomic factors most decisive. Post-communist evidence often reveals that countries with comparable levels of economic development have adopted markedly different policies toward private education and also highlights the significance of political and other factors.
7. In 1989, native Estonians constituted around 60 percent of the country's total population while the majority of non-Estonians were Russians (around 30 percent). Latvia, where ethnic Latvians were barely in majority (52 percent), had about 34 percent of the Russian population. Lithuania was an ethnically more homogenous country with almost 80 percent native and 9 percent Russian population (Lieven [1994] and Raun [1994b] in Henderson and Robinson [1997, pp. 54]).
8. It is important to note that although the 1991 Lithuanian Law on Higher Education did not address the question of private institutions explicitly, it neither proscribed their existence. Thus, attempts to open a private establishment were disapproved not on the legal but on other grounds.
9. By 2001, the sector included nine more nonuniversity private colleges that were secular (*Higher Education in Lithuania* 2001).
10. Currently, after the institutional integration process, the church-owned sector encompasses 26 institutions. Most of the theological training focuses on Christian—that is Roman Catholic and Calvinist—learning. Alongside the institutions of the Christian religion, there is the Jewish Theological Seminary

and the Gate of Dharma Buddhist College that provide Jewish and Buddhist studies respectively. It should be noted that religious institutions are mostly established as universities while nonreligious institutions, or as they are called foundation colleges, have nonuniversity status. The German Andrassy University and the Central European University, that has recently received Hungarian accreditation, are two exceptions.

11. The share of Orthodox Christians in Georgia is 65 percent.
12. However, there are borderline cases as well. Perhaps it would be still more appropriate to categorize The New Bulgarian University in Sofia, The European Humanities University in Minsk and Estonian Institutive of Humanities in Tallinn as elite institutions.
13. Correspondingly, in countries like the Czech Republic and Lithuania where significant legal barriers had existed against the private sector development, policies have been modified to allow privatization of some sort but they have remained by and large restraining.
14. The conflict that exists between public and private sectors in Georgian higher education usually intensifies before entrance examinations and manifests itself in mutual assaults by the two sector representatives. These pre-examination debates serve as a good illustration of the fact that attempts to do away with academic dishonesty and nepotism constitute one of the strongest legitimizing factors for Georgia's emergent private institutions. Private institutions are commonly blamed for their commitment to the "bottom line" and for the failings in academic standards. Revealingly, private sector representatives base their countercriticism almost solely on the overwhelming level of corruption in the public sector.
15. It should be stressed again that this concerns not private institutions in general but only those that strive for prestige and compete with traditional institutions. Otherwise, both in Ukraine and Georgia, one of the main areas of corruption identified are licensing and accreditation in which private institutions too pay bribes to obtain an authorization to operate.
16. On challenges of private higher education growth patterns to the premises of the new institutionalism see Daniel C. Levy (2004) The New Institutionalism: Mismatches with Private Higher Education's Global Growth.
17. According to Kitchelt's classification of the modes of communist rule (1999), all Soviet Union republics with the exception of the three Baltic countries and together with Albania, Bulgaria, Macedonia and Romania fall into the category of *patrimonial communism*, which distinguished itself by low levels of formal professional bureaucratization, high levels of corruption and extensive patronage and clientelist networks. The main features of *national-accommodative* communist regime were intermediate levels of formal professional bureaucratization and low-medium corruption. The latter of communism prevailed in Hungary, Slovenia, and Croatia and to a lesser extent in three Baltic republics, Slovakia and Serbia where there were also elements of patrimonial communism. Finally, *Bureaucratic-authoritarian* communism was characterized by high level of formal and professional bureaucratization, low corruption, and is largely a description of the systems in the Czech Republic, the German Democratic Republic and Poland though in the latter, elements of *national-accommodative* communism were also strong (Kitchelt 1999).

18. The interpretation of these figures should be approached with caution since public institutes do not keep a neat employment record as private institutes do.
19. It also brings to the fore one apparent contradiction that degrees granted by the "elite" private institutions enjoy much higher labor market recognition even if the public sector retains its greater prestige and status within the population at large, as elsewhere in the region.

Bibliography

Altbach, P.G. 2005. Academic Corruption: The Continuing Challenge. *International Higher Education*, No. 38, at http://www.bc.edu/bc_org/avp/soe/cihe/newsletter/News38/text003.htm, last accessed September 29, 2006.

Corso, M. 2005. Education Reform Rocks Georgia, at http://www.eurasianet.org/departments/civilsociety/articles/eav041305.shtml, accessed April 15, 2005.

DiMaggio, P. and Powell, W. 1983. The Iron Cage Revisited: Institutional Isomorphism and Collective Rationality in Organizational Fields. *American Sociological Review*, Vol. 48, pp. 147–160.

Geiger, R.L. 1986. *Private Sectors in Higher Education: Structure, Function and Change in Eight Countries*. Ann Arbor: The University of Michigan Press.

Gvishiani, N. and Chapman, D. 2002. *Republic of Georgia: Higher Education Sector Study*. Washington DC: The World Bank.

Hansmann, H. 1987. Economic Theories of Nonprofit Organizations. In Walter W. Powell (ed.), *The Nonprofit Sector*. New Haven: Yale University Press, pp. 27–42.

Henderson, K. and Robinson, N. 1997. *Post-Communist Politics: an Introduction*. London: Prentice Hall.

Higher Education in Lithuania. 2001. Vilnius: Ministry of Education.

Janashia, N. 2004. Corruption in Georgia. *International Higher Education*, at http://www.bc.edu/bc_org/avp/soe/cihe/newsletter/News34/text006.htm, last accessed September 29, 2006.

Kandelaki, G. 2005. Education: A Drag on Georgia's Reforms, at http://www.eurasianet.org/departments/civilsociety/articles/eav041405.shtml, accessed April 15, 2005.

Kitchelt, H. 1995. Formation of Party Cleavages in Post-communist Democracies: Theoretical Propositions. *Party Politics*, Vol. 1, No. 4, pp. 447–472.

Kitschelt, H., Mansfeldova, Z., Markowski, R. and Toka, G. 1999. *Post-Communist Party Systems: Competition, Representation and Inter-Party Cooperation*. Cambridge: Cambridge University Press.

Levy, D.C. 1986. *Higher Education and the State in Latin America: Private Challenges to Public Dominance*. Chicago and London: The University of Chicago Press.

——— 1999. When Private Higher Education Does Not Bring Organizational Diversity. In P.G. Altbach (ed.), *Private Prometheus: Private Higher Education in the 21st Century*. Westport: Greenwood Press, pp. 15–43.

——— 2002. Unanticipated Development: Perspectives on Private Higher Education's Emerging Roles. A PROPHE Working Paper #1, at http://www.albany.edu/prophe/publication/paper.html, accessed December 9, 2002.

——— 2004. The New Institutionalism: Mismatches with Private Higher Education Global Growth. A PROPHE Working Paper # 3, at http://www.albany.edu/dept/eaps/prophe/publication/InstitutionalismWP3.htm, accessed February 3, 2004.

Ministry of Education. 2002. *Higher Education in Hungary: Heading for the Third Millennium*. Budapest: Ministry of Education.

Mockiene, B. 2001. Transitional Higher Education in Lithuania. *International Higher Education*, 24 at http://www.bc.edu/bc_org/avp/soe/cihe/newsletter/News24/text004.htm, last accessed September 29, 2006.

Nagy-Darvas, J. and Darvas, P. 1999. Private Higher Education in Hungary: The Market Influences the University. In P.G. Altbach (ed.), *Private Prometheus: Private Higher Education and Development in the 21st Century*. Westport: Greenwood Press, pp. 161–180.

Reisz, R. 2001. Romanian Private Higher Education: A Descriptive Report, at http://www.prophecee.net/research.htm, accessed February 9, 2004.

Ruch, R.S. 2001. *Higher Ed, Inc. The Rise of the For-profit University*. Baltimore: The John Hopkins University Press.

Scott, R.W. 2001. *Institutions and Organizations*. Thousand Oaks: Sage Publications.

Sharvashidze, G. 2002. Private Higher Education in Georgia: Main Tendencies. Case study carried out under the IIEP research project on *Structural reforms in Higher Education: Private Higher Education*.

Stark, D. and Bruszt, L. 1998. *Postsocialist Pathways: Transforming Politics and Property in East Central Europe*. Cambridge: Cambridge University Press.

State Department for Statistics of Georgia. 2004. *Statistical Abstract*. Tbilisi: Ministry of Economic Development.

Stepan, A. 1994. When Democracy and the Nation-state are Competing Logics: Reflections on Estonia. *Archives Européennes de Sociologie* Vol. 35, No. 1., pp. 127–141.

Stetar, P. Panych, O. and Cheng, B. 2005. Confronting Corruption: Ukrainian Private Higher Education. *International Higher Education*, Vol. 38, at http://www.bc.edu/bc_org/avp/soe/cihe/newsletter/News38/text010.htm, last accessed September 29, 2006.

Tomusk, V. 2003. The War of Institutions, Episode I: the Rise, and the Rise of Private Higher Education in Eastern Europe. *Higher Education Policy*, No. 16, pp. 1–26.

Chapter Four

Gaining Legitimacy: A Continuum on the Attainment of Recognition

Hans C. Giesecke

Introduction

As private higher education institutions and offerings have flourished in Central and Eastern Europe over the last 15 years,[1] many questions about their legitimacy, acceptance, and trustworthiness to deliver effective instructional programs have been raised. A previous study by this author (1999)[2] underscored several key stages that many institutions of higher education travel on the road to achieving acceptance, public trust, and eventually legitimacy. Indeed, a correlative analysis incorporated in the 1999 article revealed that there is an identifiable continuum along the path to legitimation extending from institutional *effectiveness* to institutional *viability* to institutional *legitimacy*.[3]

This continuum is not envisioned as a deterministic causal chain, but rather as an "oft-trodden path" that allows a new institution to achieve its legitimacy ambitions expeditiously. Certainly, there are many ways for new private institutions of higher education to move along the road to legitimacy. Such variation in evolutionary processes reveals that the route is far from unilinear for most new institutions of higher education in the Central and Eastern European (CEE) region, yet there are patterns that repeat themselves with regularity.

The aim of this chapter is to examine the factors that affect the perceived movement of newly founded institutions of private higher education in CEE countries along this legitimacy continuum. Movement along this developmental path often has been hampered by both intentional and unintentional national higher education policies (or lack thereof) and by the less than positive perceptions of higher education authorities, policy makers, and opinion leaders. Assessing these impediments to the attainment of legitimacy enables private sector rectors and others interested in the rapid growth of this new class of institutions to consider those factors that are of the greatest significance in moving institutions forward along the given continuum.[4]

Through a web-based perceptional survey administered in May 2004 and by distilling the observations of hundreds of students[5] from the CEE region, a rank-ordered listing of the most important legitimacy factors in both Hungary and Poland was formulated. The factors delineated in this listing, while not directly applicable to the higher education circumstances of every CEE nation, orient one to those performance indicators that are likely to enable institutions to move through the legitimacy continuum at the greatest tempo. The operational supposition is that those newly founded private institutions that move through the legitimacy continuum fastest are the ones most likely to have bright futures in terms of being perceived as viable and legitimate institutions.

The perceptional survey includes 20 factors that encompass performance indicators that are frequently used by higher education accrediting agencies and ranking services in a variety of Western nations. These performance indicators were assessed for their validity by comparing them to an analysis of the development of International University Bremen (IUB) in Germany. IUB is used as an illustration because it also is a new European institution developed during the same timeframe as many of the new private institutions that were surveyed in Hungary and Poland.

Since the recent development of private higher education in Germany now mirrors in a more restrained way the phenomenal growth of private higher education in the CEE region,[6] the case of IUB is an interesting comparative example that enables a transnational crosscheck of the validity of these legitimacy indicators. A key focus of the survey was to assess those factors that may be viewed as most essential to the attainment of legitimacy in Poland and Hungary. Insights gleaned from IUB's development serve as both a crosscheck and guide to this process.

Away from the State and Toward the Market

At its most fundamental level, the privatization movement in the CEE nations reflects a move away from the state and toward the market.[7] A new breed of academic entrepreneurs has jumped into the marketplace to provide course and degree options that meet the needs and wants of students and employers through privately administered and funded alternatives. Rapid growth, however, has raised many questions about legitimacy, purpose, and function. Many observers see a class of institutions emerging that is so market-driven so as to be perceived as weak in both quality of instruction and institutional effectiveness.[8] Furthermore, perceptions of weakness have often hampered the ability of these institutions to attain their desired levels of legitimacy.

Successful institutions (i.e., those exhibiting staying power and providing evidence of generating preferred outcomes) often move through a *legitimacy* continuum with these stages:

1. Demonstration of effectiveness
2. Evidence of viability
3. Recognition of legitimate functions and outcomes

Shown diagrammatically, the continuum appears like this:

Effectiveness	Viability	Legitimacy
(evoking acceptance)	(evoking public trust)	(evoking broader recognition)

A composite theory of organizational development that hinges on an organic definition of effectiveness is utilized as the guiding paradigm for this continuum. This definition, posited by Seashore and Yuchtman (1967),[9] reads, "The effectiveness of an organization is its ability to exploit its environment in the acquisition of scarce and valued resources to sustain its functioning." This definition is particularly suitable to new private higher education institutions because it is reflective of their most pressing needs in the early stages of their lifecycles. Under this definition, institutional effectiveness is not so much a function of the extent to which an entity reflects the qualities of an ideal type, but rather, it depends much more on the match or fit between an organization's profile and its environmental circumstances.[10]

The continuum extending from *effectiveness* to *viability* to *legitimacy* may thus be viewed as a barometer of *prevailing public sentiments* that, in turn, impacts enrollment levels, institutional financial stability, and the ability to offer even more attractive and higher quality instructional programs. This analysis carries the premise that demonstration of effectiveness can and should evoke broader acceptance of private higher education entities in CEE countries among state authorities, policymakers and opinion leaders. Evidence of financial and operational viability, in turn, can and should evoke greater public trust in the institution's ability to sustain its instructional programs. Ultimately, establishment of an aura of legitimacy (or enhanced presence) through demonstrations of both effectiveness and viability should lead to greater public recognition that the outcomes and institutional products of the private sector are essential components of higher education offerings in a given country's tertiary marketplace. Growth cycle analyses of higher education entities in CEE countries support the contention that more resources used in pursuit of mission (i.e. viability) frequently engender a much more powerful image or status (i.e. legitimacy), so there is a sense of correlation between these two constructs.

The phenomenon of rapid growth in European private higher education spotlights a number of classic privatization characteristics. The foremost of these is the offering of curricular alternatives to those provided by state-sponsored institutions along with greater differentiation of institutional services. The private sector also often seeks to encourage innovations in delivery that promote efficient pursuit and completion of degrees and expeditious entry into the workforce. By being highly responsive and adaptive to student needs and concerns, these institutions frequently take the position that students are customers as well as consumers of higher education. This attitude is perceived to be necessary because students often "vote with their feet" and will change their place of enrollment if they perceive that their needs or expectations are not being met.

Rational Choice and Selection of Institutions

Students' decisions to attend private institutions of higher education in CEE countries in such substantial numbers may be explained in part by rational choice theory. This theory holds that individuals are motivated by the goals, outcomes, and wants that express their preferences. The contention is that individual actors anticipate the outcomes associated with alternative courses of action and calculate all of the likely costs and benefits of any action (or as many courses of action as they can conceptualize) before deciding which course to pursue.[11]

Within the realm of higher education, the theory postulates that prospective students will choose the higher education enrollment option(s) that is (are) likely to give them the greatest satisfaction on a variety of anticipated outcome measures. When these decisions are multiplied by thousands of students over time they begin to signal nationwide shifts in higher education enrollment demand. Such shifts, in turn, reinforce the paramount place of consumer choice in affecting higher education options; underscoring the contention that the desire to have the best possible postsecondary experience is a key driver of students' enrollment decisions and, thereby, also influential on national higher education tendencies.

Understanding why students make the institutional choices that they do is critical to an analysis of growth, effectiveness, viability, and legitimacy of the private sectors of higher education in CEE countries. A fundamental point to consider is that personal preferences, when multiplied exponentially, form the basis for thousands of institutional choices that, in turn, influence both viability and legitimacy of new private higher education entities. This is because without a continual inflow of students to stimulate their enrollments these institutions will exhibit neither viability nor legitimacy. Moreover, when a student's experience at a higher education institution is positive, his/her institutional choice is affirmed. This often leads to further evidence of successful and effective educational practices which, in turn, may result in greater enrollments, more viability, and ultimately an enhanced perception of institutional legitimacy in the higher education marketplace.

An understanding of student choices made through such rational processes must then be combined with an analysis of prevailing public sentiments regarding private higher education institutions to determine how those institutions desiring more legitimacy may attain it more readily by moving through the continuum described above. This combined approach enables one to assess the process of institutional choice through both micro and macro lenses.

Perceptions of Private Higher Education Reveal the Need for a New Private Sector Aura

Perceptions of private higher education institutions in CEE countries exhibited by state authorities, policy makers, and opinion leaders are often fraught with suspicion, mistrust, and negativism.[12] These survey findings were corroborated by hundreds of interviews conducted between fall 1997 and July 2002 with both students and administrators at new private institutions of higher education in CEE countries. Interviews underscored the perceptions that students at most private institutions of higher education are thought to be inferior; that instructional programs are considered weak and too focused on the whimsies of the marketplace; that library resources are almost nonexistent; and that the vast majority of faculty members at the new private institutions of higher education do not engage in scholarly research in their fields of inquiry (or at least when they are in the employ of a private sector entity).

Creating a greater sense of presence, however, is very difficult without demonstrating some inherent star quality that makes an institution's brand marquee shine brightly. Evidence of this phenomenon is seen in the fact that sales of commercial higher education rankings indicate a close relationship between star quality and legitimacy in the eyes of the higher education consumer. There is also substantiation of the fact that higher education authorities often rely on prevailing public sentiments created by such rankings to shape their own understanding of whether an institution is fulfilling its stated mission.[13]

Analysis of the most important legitimacy factors therefore enables sharper focus on those attributes of legitimacy that serve to solidify the presence and stature of new private postsecondary entities. Capturing the rectors' perceptions of their own stature through a web-based survey instrument administered in May 2004 allows one to assess these factors that they view as most important in gaining legitimacy in Hungary and Poland. The rectors were chosen as the target group of the study because they were the respondents on the earlier study cited above (1999) and email addresses for them were readily available through inspection of their institutions' websites. Moreover, since the rectors of these private postsecondary entities are very close to the forces and processes that shape their institutions' development, they are able to comment on these with alacrity and insight. The perceptions of the rectors, therefore, form a collage that reflects one vantage point's view of private higher education development.

Since the primary focus of the investigation was to examine the forces and factors that affect the perceived movement of newly founded institutions of private higher education along the given legitimacy continuum, an examination of prevailing perceptions is the axis around which the study revolves. A secondary thrust of the survey was to measure the extent to which, rectors in these countries believed that their institutions had been accepted as legitimate by various higher education authorities. These authorities included ministries of education, accrediting bodies, major political parties, leading in-country state-sponsored universities, and major nongovernmental organizations.

A third purpose of the survey was to determine how perceived levels of acceptance and trust exhibited by these state authorities factor into the prevailing public sentiment exhibited toward private higher education entities. A corollary activity was then to determine if country-specific means of surmounting perceived barriers to legitimacy could be identified.

Development of a Rank Ordered List of Most Important Higher Education Legitimacy Factors

The 20 factors listed in the survey were chosen initially because they are commonly used by accrediting agencies or ranking services to assess both institutional performance and status. Such factors are also used in some countries (notably the United States, Canada, and the United Kingdom) to create categorical typologies of institutions such as the Carnegie Classification System in the United States[14] or the *League Tables* of the *Times of London Higher Education Supplement.*

Survey respondents were asked to rate each of the 20 given factors on a scale of 1–5 to display their perception of each factor's importance.[15] The

20 factors that were evaluated by the Hungarian and Polish private sector rectors and rated for their relative importance as measures of institutional legitimacy were the following:

1. Amount of financial resources (budget) available to pursue stated mission
2. Date of founding
3. Official recognition by the National Ministry of Education in your country
4. Accreditation by national or international accrediting bodies
5. Professional credentials of rector and deans
6. Credentials and reputations of institutional governing board members
7. Amounts paid for administrative and faculty salaries
8. Demand by students for study places at your institution
9. Overall level of student enrollment
10. Offering of a diverse academic curriculum in five or more fields of study
11. Granting higher level degrees (e.g., Magister, Doctor, and/or Ph.D.)
12. Extent of scientific research conducted by faculty members
13. Number of articles published by faculty members in academic journals
14. Acceptance of academic credits earned by your students at other institutions
15. Affiliations with other institutions of higher education or other academic entities
16. Sponsorship by a religious denomination or church
17. Level of tuition and fees charged
18. Receipt of financial support through governmental channels
19. Contributions to your institution from the business and corporate sector
20. Recognition of your institution in national higher education rankings conducted by the media.

Results derived from this web-based survey distributed to the private sector rectors in Hungary and Poland were then used to reorder the list above according to greatest levels of significance for Poland and Hungary individually.

In Poland the top five indicators of institutional legitimacy (among those 20 factors listed above) were:

1. #20—Recognition of your institution in national higher education rankings conducted by the media.
2. #8—Demand by students for study places at your institution.
3. #9—Overall level of student enrollment.
4. #5—Professional credentials of rectors and deans.
5. #3—Official recognition by the national ministry of education.

In Hungary the top five indicators of institutional legitimacy (among those 20 factors listed above) were:

1. #4—Accreditation by national or international accrediting bodies.
2. #3—Official recognition by the national ministry of education.
3. #11— Granting of higher level degrees (e.g., Magister, Doctor, and/or Ph.D.).
4. #5—Professional credentials of rectors and deans.
5. #1—Amount of financial resources (budget) available to pursue stated mission (tie with 15).
6. #15—Affiliations with other institutions of higher education or other academic entities.

(**Note:** There was a tie with #1 and #15 for 5th place.)

Differences between Poland and Hungary in the top five indicators of institutional legitimacy reflect differences in the national circumstances of recent private higher education development in these two countries. According to the Polish rectors, private sector legitimacy is reflected primarily by demand for enrollment/study places at various institutions. If demand is high, it means that it is a legitimate place for students to study.

The major factor that affects such demand is the national ranking of higher education institutions conducted by the educational publishing house *Perspektywy* in conjunction with a leading Polish newspaper *Rzeczpospolita*. Those two entities combined forces in 1992 to produce the first commercially available education ranking table in the country and have since prepared and distributed a dozen editions of this ranking.[16] These institutions that fare well in the rankings boldly advertise their results across their web pages and in nearly every publication they produce. It is clear that in Poland the broad utilization and discussion of these rankings appears to fuel demand for study places at certain institutions in significant ways.

In Hungary, private sector legitimacy hinges more heavily on the receipt of accreditation from the national ministry of education's accrediting council. Once this accreditation is received, private sector institutions gain some access to public funding and to a system of student loans that supports enrollment at private sector institutions. Without such accreditation, an institution has little chance of success in the Hungarian higher education marketplace. By attaining such accreditation, a private sector institution is able to achieve almost equal standing with state-sponsored universities and make bold pronouncements about the quality of instruction it offers within the context of the higher education market that it serves.

A major difference between Poland and Hungary in terms of gaining legitimacy is reflected in their varying approaches to higher education access. Poland's private sector market share of approximately 30 percent is more than twice the market share of Hungary's 14 percent. The Polish case shows heavy dependency on enrollment for gaining legitimacy largely due to student demand and the creation of academic programs that meet such demand. In contrast, the route to gaining legitimacy in Hungary is much more dependent on meeting government-imposed standards—even if these institutions' leaders do not actually accept these criteria as true measures of their legitimacy.

Perceptions of Institutional Acceptance and Sense of Trust in Being Treated Fairly

Polish and Hungarian private sector rectors were also queried on their perceptions of fair treatment and the measure of acceptance given to their institutions by higher education authorities. Two questions were posed to help the rectors identify the major sources of resistance to their quest for legitimacy in their specific national and cultural contexts. These questions were designed to elicit responses that would help the rectors craft approaches that they might use to convert the perceptions of their most visible detractors from negatives to positives.

1. How well have private (nonstate) institutions of higher education been accepted as legitimate institutions by the various higher education authorities in your country?
2. What level of trust do you have that your institution is being treated fairly by higher education authorities in your country?

According to the private sector rectors in Poland, the *lowest* levels of acceptance were perceived to be from:

1. Leading state-sponsored universities
2. Major political parties in-country
3. National and/or international accrediting bodies,
4. Major nongovernmental organization in-country
5. National Ministry of Education

Thus, in Poland the lowest level of acceptance was perceived to be from the state-sponsored universities and the highest level of acceptance from the National Ministry of Education.

According to the private sector rectors in Hungary, the *lowest* levels of acceptance were perceived to be from:

1. Leading state-sponsored universities,
2. National Ministry of Education,
3. Major political parties in-country,
4. National and/or international accrediting bodies,
5. Major nongovernmental organizations in-country.

Thus, in Hungary the lowest level of acceptance was perceived to be from the state-sponsored universities and the highest level of acceptance from major nongovernmental organizations in-country.

Furthermore, among the Polish private sector rectors, the *lowest* level of perceptional trust that their institutions are being treated fairly stemmed from:

1. Major political parties in-country
2. Leading state-sponsored universities
3. National and/or international accrediting bodies
4. Major nongovernmental organization in-country, then from
5. National Ministry of Education

In Poland the lowest level of perceptional trust in being treated fairly stemmed from major political parties and the highest level of perceptional trust in being treated fairly stemmed from the National Ministry of Education. However, it should be noted that the range of scores was very close and that all were weighted more toward lower levels of perceptional trust.

Among the Hungarian private sector rectors, the *lowest* level of perceptional trust that their institutions are being treated fairly stemmed from:

1. National and/or international accrediting bodies
2. Leading state-sponsored universities
3. National Ministry of Education
4. Major political parties in-country
5. Major nongovernmental organizations in-country

Thus, in Hungary the lowest level of perceptional trust in being treated fairly stemmed from national and/or international accrediting bodies and the highest level of perceptional trust in being treated fairly stemmed from major nongovernmental organizations in-country.

High levels of correspondence between the entities perceived to offer the lowest levels of acceptance and trust in Poland and Hungary underscore the point that these two constructs are closely correlated. Since state-sponsored universities offer the perceived lowest levels of acceptance in both countries and the second lowest-levels of perceived trust in being treated fairly in both countries, they appear to be the most significant detractors of the private sector growth and development in these countries.

This finding is perplexing because most faculty members at private sector institutions in Poland and Hungary either come from or are concurrently employed by state-sponsored universities. A possible explanation is that it is the structure, purpose, and function of private higher education, rather than the personnel involved, that often leads it to be held in such low regard by leaders of state-sponsored universities in both Poland and Hungary.

Perceived lack of acceptance and mistrust by political parties is also a compelling finding of the survey responses submitted by private sector rectors in both Poland and Hungary. Because members of the national accrediting councils are appointed by the ruling political parties, their policies/directives are often controlled by leaders of state-sponsored universities who also frequently have close party ties. This is an explanation of why there has been such suspicion among private sector rectors in Hungary for the actions of the national higher education accrediting council.

These perceptional findings are reflective of one of the major challenges facing private sector rectors in Central and Eastern Europe. If private sector institutions are beholden to the national accrediting councils that are appointed largely by the ruling political parties and perceived to be unduly influenced by state-sponsored university leaders, then they must learn to play the game of politics very adeptly. This is the case because landing on the "outs" with a ruling political party might cost an institution its legal standing, and hence, its ability to attract students and survive in the higher education marketplace.

Legitimacy Factors in Regional Comparison: The Case of IUB in Germany[17]

One way to gain a better sense of whether those indicators of legitimacy rated highest by the private sector rectors in Hungary and Poland are indeed valid cross-national measures is to view them in the context of a new private institution in a neighboring country. The case study of IUB offers an evocative illustration of how the legitimacy continuum actually functions in a setting where the results have been both observable and verifiable.

IUB was founded on the outskirts of the port city of Bremen in Northwestern Germany in February 1999 largely to emulate the economic development model developed by the Massachusetts Institute of Technology in Boston (USA) and by the California Institute of Technology in Southern California (USA). With unemployment in the Bremen region approaching 20 percent at the end of the 1990s due to cutbacks in the region's shipbuilding industry, leaders of the city-state were seeking a means of making a strategic investment to spur on economic growth and reduce the unemployment rate. They seized on the idea of establishing a new private university to accomplish this objective. Since IUB moved through the legitimacy continuum with great speed after its founding, one can identify the key stages along the way that were critical to its move toward legitimation.[18] The example of this new private university in Germany sheds light on the steps that new private institutions in Central and Eastern Europe can take to achieve more recognition within their own national and cultural contexts.

While most of the 55 plus new private institutions of higher education now operating in Germany are highly specialized institutes focusing on only one or two subject fields, IUB was deliberately planned to be a comprehensive research university with an international focus and first-rate science and engineering offerings.[19] The university had four major operational elements in place shortly after its founding. First, it had possession of a well-developed 30 hectare (75 acre) campus that was a former officer's training academy of the German army. Second, it had an endowment of around 115 million Euros that was granted from the city-state Bremen using federal economic development funds as founding capital to get the project off the ground. Third, it had a solid educational alliance with Rice University in Houston, Texas United States. And fourth, it had excellent political support from both the mayor and city-state's senate.[20]

While IUB's birth advantages made its founding circumstances quite different from most of the then new private institutions of higher education in Poland and Hungary, it also shares a number of attributes with them. As a new institution, IUB has had to go through many of the same stages of the legitimacy seeking process that the new institutions of Central and Eastern Europe have been compelled to transit. As a privately governed institution, IUB must charge tuition fees and seek much of its funding from non-governmental sources. Having no prestige or identity of its own, the university has had to create its own brand image. And, since it is located in a neighboring European country, IUB's development is being impacted by the same European-wide higher education mandates that are impacting institutions across Central and Eastern Europe.

IUB's effort to enter the European higher education market as an elite institution also makes its scope of mission and operational circumstances different from most of the new private institutions of higher education in both Poland and Hungary. Yet, this quest for elite status has created environmental circumstances that have enabled it to test the legitimacy continuum paradigm in a very public and visible way. This is because every step of IUB's development has been closely inspected by the media and heavily reported in the German and pan-European press.[21] Another reason is that IUB's creation came about largely because there was a national outcry in Germany that there were no German universities (public or private) listed among the world's top 50 universities.[22] IUB's quick development was partially a result of a perceived lack of quality, flexibility, and responsiveness in the German state-sponsored university system. These are classic arguments for privatization, as they connote a breakup of state monopoly to provide better service to students and create more effective higher education outcomes.

As IUB moved through the legitimacy continuum, it readily achieved several milestones that enabled it to surmount barriers along the continuum. While it had been granted state accreditation—*legal authority to operate*—by the city-state Bremen shortly after its founding, the university needed a more well-recognized seal of approval that would endow it with the stamp of authority needed to attract research funding and the confidence of private donors. This seal of approval was granted when IUB became the second private institution in Germany (following the university of Witten-Herdecke) to receive the imprimatur of the *Deutsche Wissenschaftsrat* (The German Scientific Council).

Then, in June 2004, IUB had all 18 of its bachelor's degree programs accredited by the agency known as ACQUIN, the German Accreditation, Certification and Quality Assurance Institute, which is a nongovernmental agency with an independent stance toward academic standards and traditions. Accreditation of the various bachelor's degree programs gave IUB's offerings even more credibility in that, they were now considered officially comparable to those of other European universities and warranted as being of the quality and type agreed to by the Bologna Declaration of 1999. Although not required by law or even by student market demand, this supplemental accreditation of IUB's individual academic programs gave the university the ability to extol its perceived quality in the media and other higher education circles.

The question of tuition pricing at IUB was a key concern. After significant study and debate, the university eventually settled on an annual tuition fee of 15,000 Euros per academic year. This was an exceptionally high rate by European standards and it became part of the brand image of the university. To convince a skeptical public that the university was not only for the financially well situated, IUB announced in all of its public documents and press releases that it would be need-blind in terms of its admission policy and that generous financial aid packages would be offered to students who were admitted but could not afford the tuition charge.

The offering of scholarships became one of the central components of the student marketing campaign. The university's team of student admissions and outreach people made the availability of generous financial aid and scholarship awards one of its top selling points. Through the charging of substantial tuition fees (with commensurate reduction of such tuition through scholarship aid), IUB further enhanced its aura of legitimacy. This was the outcome, because when consumers express—as a manifestation of *rational choice*—a willingness to expend their own resources for a commodity or service, they validate its usefulness and worth to them personally. IUB, therefore, provided more higher education access and choice to the public through its financial aid and curricular offerings yet also gained legitimacy by asking individual beneficiaries to pay what they could in tuition fees to demonstrate their assignment of value to the educational services they received.

Despite IUB's success in promoting enrollment growth since the arrival of its first students in fall 2001,[23] there has been an ongoing apprehension about students' willingness to pay tuition fees in a national context where students enrolled at state-sponsored universities pay virtually nothing. Even with a comprehensive program of student financial assistance in place, there has been some doubt that German students, in particular, would enroll at IUB in substantial numbers with such a sizeable tuition gap between it and state-sponsored universities.

Intense focus on attainment of enrollment goals resulted in a strong display of demand among international students for this new private, English-language institution on the European continent. Indeed, most of the best and brightest students in the university's initial entering cohorts came from former Warsaw Pact countries where they had been well-schooled in mathematics and the sciences and learned to read and write well in English.

Now in its fifth year of instructional operations (2005/2006), IUB has enrolled some 930 students at both the undergraduate and graduate levels hailing from more than 80 countries around the world.[24] More than half of the student body, however, still comes from post-communist countries in Central and Eastern Europe with Bulgaria, Romania, Poland, Lithuania, and Hungary leading the way as the "Big Five" in terms of student nationalities from this region. It is clear that students from these countries remain dissatisfied with the more traditional enrollment opportunities at state-sponsored universities within their own countries and, therefore, have enrolled at IUB in substantial numbers. This phenomenon explains, to a great degree, the massive expansion of in-country private institutions in Central and Eastern Europe. For, if the state-sponsored universities had offered the heterogeneity and volume of curricular options that the private sector now presents, then one might hypothesize that private sector expansion would have been far less than it has actually been.

Table 4.1 Legitimacy Factors with Combined Mean Scores Attributed by Rectors in Poland and Hungary

(1-High, 5-Low)		
Mean Rating	*Legitimacy Factor*	*Present/Absent at IUB*
2.05	Recognition of your institution in national higher education rankings	Yes*
2.09	Official recognition by the National Ministry of Education in Your Country	Yes
2.10	Accreditation by national or international accrediting bodies	Yes
2.23	Demand by students for study places at your institution	Yes—Int'l **
2.23	Professional credentials of rectors and deans	Yes
2.41	Overall level of student enrollment	No
2.55	Granting higher level degrees (e.g., Magister, Doctor, and/or Ph.D.)	Yes
2.55	Affiliations with other institutions of higher education or other academic entities	Yes
2.62	Amount of financial resources (budget) available to pursue stated mission	Yes
2.73	Number of articles published by faculty members in academic journals	Yes
2.82	Offering of a diverse academic curriculum in five or more fields of study	Yes
2.86	Extent of scientific research conducted by faculty members	Yes
2.90	Date of founding	No
2.91	Contributions to your institution from the business and corporate sector	Yes
2.95	Credentials and reputations of institutional governing board members	Yes
3.09	Acceptance of academic credits earned by your students at other institutions	Yes
3.14	Amounts paid for administrative and faculty salaries	Yes
3.14	Receipts of financial support through governmental channels	Yes
3.23	Levels of tuition and fees charged	Yes
4.05	Sponsorship by a religious denomination or church	No

*Notes** IUB has only been listed in the higher education ranking system for Germany developed by the Center for Higher Education Development (CHE) in conjunction with the German magazines *Der Stern* and *Die Zeit* since May 2005.

** IUB's demand for enrollment has stemmed primarily from non-German students, especially those from postcommunist countries.

Legitimacy Factors from Poland and Hungary Known to be Present/Absent at IUB

Description of IUB's development and movement through the legitimacy continuum allows an analysis of the key combined legitimacy factors that were rated as most important by the Polish and Hungarian private sector rectors. Table 4.1 shows the 20 indicators of legitimacy ranked according to their level of importance assigned them by the private sector rectors in both Hungary and Poland. It is shown alongside the presence/absence analysis of indicators at IUB to reflect how these factors also are descriptors of the process of gaining legitimacy.

Observations About Gaining Legitimacy

A number of observations can be drawn from the comparison of key legitimacy factors that are present/absent at IUB in Germany and new private institutions in Poland and Hungary. Establishing a presence of 17 out of 20 of these factors in less than five years of instructional operations explains in part why there has been so much media attention given to this particular university development project.[25] Its relatively late appearance (since May 2005) in the German higher education rankings developed by the CHE apparently has not affected overall student demand for study places at IUB, although it may partially explain why the percentage of native German students at the undergraduate level has remained below 15 percent during its first four years of operation.[26] Demand for undergraduate enrollment spaces at IUB by German nationals has clearly been one of the university's few areas of weakness. While most of the German students who eventually enroll at IUB come with top secondary school credentials, there have not been nearly as many "full-pay" German undergraduates as the university's planners originally envisioned. This trend, however, appears to have been modestly reversed with the cohort entering in fall 2005.

The "willingness to pay" factor with respect to German students must be clarified to mean a "willingness to accept a scholarship and/or loan" in order to try out a different kind of higher education experience. This clarification underscores the point that IUB's international legitimacy amongst both students and educational advisers in the countries of Central and Eastern Europe has been achieved much faster than it has in other contexts. The main explanation for this is that students from Western countries have a much better set of higher education alternatives to choose from, than do students from post-communist countries and, consequently, they do not immediately see the opportunity to study at IUB as a rational choice. Studies in the field of "postsecondary choice" reveal that students make the enrollment decisions that they do in order to make the most of their various options.[27] When students from western nations with well-developed higher education options consider IUB as an enrollment possibility, they normally compare it to the other options they have in their home countries. Under such scrutiny, IUB appears excellent to some and less appealing to others depending on the individual's basis of comparison and personal preferences.

As noted in the case of IUB, it may *not* be such a negative factor to maintain enrollment levels below a benchmark level of 1,000 students when striving for differentiation in the crowded higher education marketplace. This is because a somewhat smaller and more selective student body may actually be a positive in the development of an institution's reputation; that is, especially, if operational budgets can be sustained at the desired level of enrollment and tuition revenue. Indeed, maintenance of a lower enrollment level may allow for a more personable student-faculty ratio and reports from many different countries indicate that institutions with lower levels of enrollment often have shorter time-to-degree rates—an attribute that can be a great economic efficiency advantage for both individuals and institutions.

While it will take another century or so for IUB to achieve the legitimacy factor of age maturity, the significance of the age factor was obviated by the institution's intense focus on excellence. Since the university's unofficial motto of "Think big, but stay small!" appears to be attracting sufficient financial resources in the near term, the problem of the need to increase enrollments dramatically each year to sustain revenue yields also does not appear to be overly problematic. And since the university may be listed at or near the top of most of the German higher education rankings, it can begin to lay claim to greater public recognition through this key legitimacy avenue as well.[28]

Clearly, IUB's newness is a quality that it shares with many of its sister institutions in the CEE region. Its novel and innovative character is something that bonds it together with many institutions that have been created east of the Oder-Neisse line. The clear realization is that many key characteristics of privatization—offering alternatives to those provided by the state sector, providing greater differentiation of services, and encouraging innovations in delivery that promote efficient pursuit and completion of degrees—are exhibited conspicuously by both IUB and the majority of new private institutions of Central and Eastern Europe.

The listing of most significant legitimacy factors provides a starting point from which the rectors of the private institutions of Central and Eastern Europe can assess their progress along the road to legitimacy. An assumption of 75–80 percent (15–16 out of 20) of these key legitimacy factors is an indication that a new institution is well along the continuum that starts with effectiveness and then moves to viability and then on to legitimacy.

Summary and Questions for Further Analysis and Investigation

One of the most important questions to consider in this discussion is: Are these key legitimacy factors cross-nationally transferable and generalizable? The case of IUB shows that in a limited way these are valid cross-national measures insofar as all private higher education institutions share some common characteristics and developmental stages. Since private institutions are nongovernmental entities that rely heavily on tuition fees as their major source of revenue, it is clear that they must constantly seek to deliver instructional services in a manner that encourages and incentivizes their clientele(s) to keep coming back for more. Indeed, a prominent view of organizational effectiveness emphasizes that in order to sustain their existences, private sector entities must satisfy the needs of their clients by providing adequate inducements to sustain their continued contributions.[29]

Private institutions, in a global sense, are likely to be more focused on their public personae because of such market sensitivities. When institutions achieve greater public recognition and support for the educational services they are rendering, this often becomes a test of legitimation. Such public recognition may be measured in the amount of funds raised from private donors and corporations, the creation of state-subsidized scholarship programs that benefit students, and consistent rankings among the top-tier institutions in their home countries.

It is evident that the appropriateness of legitimacy factors is heavily dependent on key characteristics of the national higher education systems where they are applied and utilized.[30] For instance, a factor may have an important meaning in one country and generate very little reaction in another cultural or national context. A good example of this is the degree of importance a rector gives the issue of tuition fees. According to the Hungarian and Polish rectors responding to the survey, the level of tuition fees charged by institutions was one of the lowest ranked legitimacy factors. In many western countries charging higher tuition fees is often taken to be a measure of quality and status as this notion undergirds the "Chivas Regal Effect" (i.e., what costs more must naturally be of better quality). For this reason, many rectors and presidents of private institutions in the United States and other Western countries stay carefully attuned to tuition rates within their peer groups and scrutinize those institutions that are acting outside peer group norms on such questions.

The case study of IUB, moreover, informs us that institutions move through the legitimacy continuum at considerably different tempos, depending on their environmental circumstances and various sources of support. When institutions are able to gain the majority of those highest ranked key legitimacy factors in a short period of time, they are also likely to attain a measure of public recognition much faster. IUB's move toward legitimacy was speeded by the fact that it received the imprimatur of the city-state where it is located at a very early stage in its development and then used this official backing to obtain a campus, an endowment, accreditation from the city-state, and ultimately the recognition of the German Scientific Council that, in turn, provided access to federal research funding.

The issue of faculty research is something that clearly plays out in disparate ways in certain countries according to their historical contexts and higher education traditions. For institutions that describe themselves as universities, no matter where they are located, research must be an important part of their mission. For institutions that call themselves "higher schools" or "institutes," which is what the majority of the new private institutions in CEE countries actually are, the pursuit of research is seldom a critical step for bolstering their legitimacy. In fact, the rankings provided by the private sector rectors in both Hungary and Poland showed that the pursuit of research by their faculty was not considered a top ten legitimacy factor.

Given that the rectors attribute great significance to the media rankings of their institutions, it is vital to take stock of the extent to which such rankings of higher education institutions artificially affect the legitimacy of certain types of institutions. According to student user data collected via survey by the CHE in Germany, about one-third use rankings for orientation about which institutions they should consider for post-secondary enrollment. Students in scientific and technical fields such as engineering tend to make use of these rankings more often than students in the humanities or social sciences. In general, it has been found that high-achieving students tend to use and make good use of the rankings more than those who are somewhat lackadaisical in their approach to higher education.[31]

If these new private institutions want to dispel the common belief that their students are inferior to those enrolling at state-sponsored institutions, they must be concerned about their positions in the rankings. There is a strong assumption that only by attaining higher rankings can they attract better students and break the magnetic force that the state-sponsored universities have developed over decades for enrolling the highest achieving students.

The question remains whether these rankings affect an institution's trajectory toward legitimacy. When constructed properly, university ranking systems focus primarily on the environmental factors and academic conditions that are already in existence at a particular institution. The thrust of this study has been to show that private institutions must first demonstrate effectiveness and viability to attain a state of legitimacy. If an institution successfully provides evidence of being effective in moving students through its instructional programs and shows that it can do so in ways that give it a stable operating base, then it has a much greater chance of being perceived as a legitimate institution. Ranking schemes that provide objective and verifiable assessments of institutional presence and service, therefore, are also likely to impact perceptions of legitimacy.

Some institutions learn out of necessity how to improve their standings by tweaking and adjusting some of their own outcome measures that go into producing the rankings. There are even reports of institutions going through contortions to improve their rankings by shifting resources and faculty from place to place so as to create better impressions in certain fields of study.[32] Moreover, there is always the temptation to focus quality improvements on the specific assessment measures that the commercial enterprise is using to develop the ranking. Because the chief motive for producing such rankings is often profit or visibility, firms may accept assessment outcomes from certain institutions that experienced higher education policy researchers would not consider. This may have a distortional effect on the rankings and give those institutions that seek to advance their positions through "creative manipulation" an upper hand in the process.

Understanding more about the sources of tension over institutional legitimacy between private and state-sponsored institutions of higher education may also aid in improving the low levels of acceptance and trust that the state-sponsored universities afford the new private institutions. If private sector rectors can determine why the leaders of state-sponsored universities often hold them in such low regard, they may implement steps to improve their relationships—which ultimately should hold many advantages for students, who are the chief beneficiaries of higher education systems that demonstrate mutual respect and compatibility. Indeed, more appreciation of the factors that lead to differentiation of mission and service should lead to greater social acceptance. As these new private institutions are able to demonstrate both effectiveness and viability, they also will be viewed as increasingly legitimate by the opinion leaders and prospective students who will largely determine their collective futures.

Notes

1. See data in the Program for Research on Private Higher Education (PROPHE) website at: http:// www.albany.edu/dept/eaps/prophe/data/PROPHEDataSummary.doc or see K. Galbraith (2003), Toward quality private higher education in central and eastern Europe, in *Higher Education in Europe*, Vol. 28, No. 4, p. 540, or see J. Woznicki and R.Z. Morawski (2002), Public and private higher education institutions—joint or separate evaluation and ranking: the Polish perspective in *Higher Education in Europe*, Vol. 28, No. 4, pp. 461–462. Comparative data also is provided in the Comparative analysis on private higher education in Europe, prepared by J. Fried, A. Glass, and B. Baumgartl (2005), a project paper published by UNESCO–CEPES for the International Conference

on Private Higher Education in Europe and Quality Assurance and Accreditation from the Perspective of the Bologna Process Objectives.

2. H.C. Giesecke (1999), The rise of private higher education in east central Europe, in *Society and Economy*, Vol. 21, No. 1, pp. 132–156.
3. Substantial evidence in support of a continuum with these stages is noted through an examination of the evaluation processes utilized by voluntary national and regional accrediting agencies. Accreditors first look to see whether an institution of higher education has achieved its stated mission by enrolling and graduating students in academic degree programs consonant with its goals and prerogatives (effectiveness). Then, accrediting agencies examine whether the institution under review has the financial wherewithal and resources necessary to sustain this mission over time (viability). Finally, accrediting committees want to verify that an institution is able to promote and substantiate the effectiveness of its educational practices for its target populations and for broader acceptance in the higher education and employment marketplaces (legitimacy). These points are spelled out in the "The Criteria for Accreditation" highlighted in the North Central Association of Schools and Colleges Higher Learning Commission document entitled, Institutional Accreditation: An Overview (2003 Edition), effective January 2005. Higher Learning Commission of the North Central Association of Colleges and Schools, Chicago, Illinois.
4. An important qualifier of the data reported in this study is that it represents perceptions and opinions. The rectors of private sector institutions in Hungary and Poland were surveyed regarding their *perceptions* of most important legitimacy factors and most significant impediments to the achievement of legitimacy. These are representations of their opinions regarding these questions and not analyses of quantifiable metrics.
5. This author conducted more than 500 personal interviews with university–bound students and campus administrators in Bulgaria, the Czech Republic, Estonia, Hungary, Latvia, Lithuania, Poland, Romania, and Slovakia between January 2000 and July 2002. The results of these interviews were reported in the document, H.C. Giesecke (2002), It's all about choices, international student marketing and outreach strategies, an unpublished report prepared for the senior administration of International University Bremen in Germany, June 2002.
6. According to the Statistical Office of the German Federal Ministry of Education and Research, there were 56 private higher education institutes and universities in Germany in 2003–04 enrolling 39,052 students. These numbers reflect nearly four–fold growth from 1992 to 93 when there were 28 private institutions enrolling 11,563 students. These data were reported in the May 2005 issue of *Abi Berufswahl-Magazin* online.
7. M. Kwiek, The Missing Link: Public policy for the private sector in central and east European higher education, in *SRHE International News* London, No. 52, Summer 2003, pp. 6–8.
8. H.C. Giesecke (1998), The rise of private higher education in East Central Europe: characteristics affecting viability and legitimacy by institutional type, Ph.D. dissertation at Vanderbilt University, UMI: Ann Arbor, MI, pp. 81–85.
9. S.E. Seashore and E. Yuchtman (1967), Factorial analysis of organizational performance. *Administrative Science Quarterly* Vol. 12, pp. 377–395.

10. K.S. Cameron and D.A. Whetten (1996) Organizational effectiveness and quality: the second generation in *Higher Education: Handbook of Theory and Research*, Vol. 11, J.C. Smart, (ed.) (Agathon Press, New York, NY), pp. 269–271.
11. J. Scott (2000), Rational Choice Theory, in G. Browning, A. Halcli, and F. Webster (eds.), *Understanding Contemporary Society: Theories of the Present*, (Sage Publications, 2000) pp. 1–2.
12. R.D. Reisz (2003), Public policy for private higher education in central and eastern Europe. *Arbeits Berichte*, No. 2 HoF Wittenberg-Institut fuer Hochschulforschung: Leucorea.
13. B. Nedwek (1996), Public policy and public trust: The use and misuse of performance indicators in higher education, in *Higher Education: Handbook of Theory and Research* Vol. 11, (Agathon Press, New York, NY), pp. 47–48.
14. A.C. McCormick and C.M. Zhao (2005) Rethinking the and Reframing the Carnegie Classification in *Change: The Magazine of Higher Learning*, September/October 2005, pp. 51–57.
15. Likert rating scale definitions used by the respondents to assess the importance of key legitimacy factor are:

    1. Highly significant legitimacy factor;
    2. Very significant legitimacy factor;
    3. Moderately significant legitimacy factor;
    4. Weak legitimacy factor;
    5. Not a factor in institutional legitimacy.

16. W. Siwinski, 2002, Perspektywy—Ten Years of Rankings, *Higher Education in Europe*, Vol. 28, No. 4, p. 399.
17. In November 2006, the Jacobs Foundation of Switzerland pledged to give more than $250-million over the next five years to IUB. Its name will be changed to Jacobs University Bremen.
18. This author served as a full–time consultant–in–residence at International University Bremen from January 2000 through July 2002 and personally witnessed the University's rapid movement through the legitimacy continuum.
19. Philipp Scheffbuch (2004). Management auf Englisch: Privat Unis in Deutschland in *Stuttgarter Zeitung*, January 9.
20. IUB Report entitled, *Committed to Excellence*, dated Summer 2001.
21. See for example, Wo studiert die elite? in *Die Zeit*, September 8, 2005, p. 40.
22. See The Academic Ranking of World Universities—2004 that was developed by Jiao Tong University's Institute of Higher Education (China) at http://ed.sjtu.edu.cn/ranking.htm, last accessed October 4, 2006.
23. IUB Student Enrollment Report, October 2004, prepared by the IUB Office of the Registrar.
24. These data were touted in the document, Greeting from the President, on the IUB website, dated June 2005 at www.iu-bremen.de/about/president/print.php, last accessed October 4, 2006.
25. See for example the article in *Time* Magazine (European Edition) dated April 8, 2002 which featured IUB prominently, entitled Germany's Ivy League: Fee-paying universities are attracting students seeking quality tuition and a fast-track degree. See also Picasso in der Bewerbungsmappe in *Financial Times*

Deutschland, April 22, 2004, p. 28 or Studieren im Schloss in *Frankfurter Allgemeine Zeitung*, February 28, 2004, p. 58.

26. Letter to the Editor written by Alexander Ziegler-Joens, IUB Vice President for Business Affairs at IUB, in the student publication *Pulse of the World*, November 15, 2004. This may be viewed at www.pulseoftheworld.com, last accessed October 4, 2006.
27. D. Hossler, J. Schmit, and N. Vesper (1999), *Going to College: How social, economic, and educational factors influence the decisions students make* (The Johns Hopkins University Press, Baltimore and London), pp. 141–155.
28. See IUB's own website reference to its initial CHE rankings at www.iu-bremen.de/news/iubnews/05612/, last accessed October 4, 2006.
29. K.S. Cameron and D.A. Whetten (1996), Organizational effectiveness and quality: the second generation, in J.C. Smart, (ed.), *Higher Education: Handbook of Theory and Research*, Vol. 11, (Agathon Press, New York, NY) pp. 269–271.
30. G. Federkeil. 2002. Some aspects of ranking methodology—The CHE ranking of German universities in *Higher Education in Europe*, Vol. 27, No. 4, p. 394.
31. Ibid., p. 395.
32. J. Vaughn. 2002. Accreditation, Commercial rankings, and new approaches to assessing the quality of university research and education programmes in the United States. *Higher Education in Europe*, Vol. 27, No. 4, p. 436.

Bibliography

Cameron, K.S. and Whetten, D.A. 1996. Organizational Effectiveness and Quality: The Second Generation. In J.C. Smart (ed.), *Higher Education: Handbook of Theory and Research*, Vol. 11, New York, NY: Agathon Press. pp. 269–271.

Federkeil, G. 2002. Some Aspects of Ranking Methodology—The CHE Ranking of German Universities. *Higher Education in Europe*, Vol. 27, No. 4, p. 394.

Fried, J., Glass, A., and Baumgartl, B. 2005. Comparative analysis on private higher education in Europe, a project paper published by UNESCO–CEPES for the International Conference on Private Higher Education in Europe and Quality Assurance and Accreditation from the Perspective of the Bologna Process Objectives. Warsaw, Poland, November.

Galbraith, K. 2003. Towards Quality Private Higher Education in Central and Eastern Europe. *Higher Education in Europe*, Vol. 28, No. 4, p. 540.

Giesecke, H.C. 2002. It's all about choices, international student marketing and outreach strategies. An unpublished report prepared for the senior administration of International University Bremen in Germany, June.

——— 1999. The Rise of Private Higher Education in East Central Europe. *Society and Economy*, Vol. 21, No. 1, pp. 132–156.

——— 1998. The Rise of Private Higher Education in East Central Europe: Characteristics Affecting Viability and Legitimacy by Institutional Type. Ph.D. dissertation at Vanderbilt University, UMI: Ann Arbor, MI, pp. 81–85.

Kwiek, M. 2003. The Missing Link: Public Policy for the Private Sector in Central and East European Higher Education. *SRHE International News* London, No. 52, Summer, pp. 6–8.

McCormick, A.C. and Zhao, C.M. 2005. Rethinking and Reframing the Carnegie Classification. *Change: The Magazine of Higher Learning*, September/October, pp. 51–57.

Nedwek, B. 1996. Public Policy and Public Trust: The Use and Misuse of Performance Indicators in Higher Education. In *Higher Education: Handbook of Theory and Research*. J. C. Smart (ed.) Vol. 11, New York, NY: Agathon Press., pp. 47–48.

Reisz, R.D. 2003. Public Policy for Private Higher Education in Central and Eastern Europe. *Arbeits Berichte*, No. 2 HoF Wittenberg–Institut fuer Hochschulforschung: Leucorea.

Scheffbuch, P. 2004. Management auf Englisch: Privat Unis in Deutschland. *Stuttgarter Zeitung*, January 9.

Scott, J. 2000. Rational Choice Theory. In G. Browning, A. Halcli, and F. Webster (eds.), *Understanding Contemporary Society: Theories of the Present.* Sage Publications Ltd.: London, U.K. pp. 1–2.

Seashore, S.E. and Yuchtman, E. 1967. Factorial Analysis of Organizational Performance. *Administrative Science Quarterly* Vol. 12, pp. 377–395.

Siwinski, W. 2002. Perspektywy—Ten Years of Rankings. *Higher Education in Europe*, Vol. 28, No. 4, p. 399.

Vaughn, J. 2002. Accreditation, Commercial Rankings, and New Approaches to Assessing the Quality of University Research and Education Programmes in the United States. *Higher Education in Europe*, Vol. 27, No. 4, p. 436.

Weber, M. 1947. The *Theory of Social and Economic Organization.* A.M. Anderson and Talcott Parsons, trans. New York: Oxford University Press. p. 153.

Woznicki, J. and Morawski, R.Z. 2002. Public and Private Higher Education Institutions—Joint or Separate Evaluation and Ranking: The Polish Perspective. *Higher Education in Europe*, Vol. 28, No. 4, pp. 461–462.

Chapter Five
The European Integration of Higher Education and the Role of Private Higher Education

Marek Kwiek

Introduction

The Bologna process—a major European integrating initiative in higher education started by the Bologna Declaration in 1999 and to be completed by 2010—seems to disregard one of the most significant recent developments in several major post-communist transition countries in Central and Eastern Europe: the rise and rapid growth of the private sectors in higher education and, more generally, the emergence of powerful market forces in higher education. Consequently, the ideas behind the Bologna process, the analytical tools, and policy recommendations it provides may have unanticipated effects on higher education systems in certain Central and Eastern European countries. The growth of both the private sector in European (and especially Central and East European) higher education systems and the emergence of powerful market forces in the educational and research landscape in Europe warrant further consideration by the Bologna process if it is not to turn into a merely "theoretical," myopic exercise. The downplaying of the role of market forces in higher education and research and development in the Bologna documents and the omission of the private sector (with its evident successes in some places and failures in other places) from the overall conceptual scheme of the Bologna process give potentially misguided signals to educational authorities in transition economies. Consequently, the Bologna process might thwart the development of the private sector in countries where chances for the expansion of the educational system otherwise than through privatization have been limited. The expansion of educational systems here is crucial for the implementation of the Lisbon strategy of the European Union, as described briefly in the next section. Thus, while the implicit disrespect for market mechanisms in higher education may have limited impact in Western European systems, which have increasingly taken many market-related parameters of their operation in public universities for granted, it might have long-lasting negative impact on legislation and general attitude toward the private sector in some Central and Eastern European countries. With its magnitude, the role of the Bologna process in (indirectly) granting or refusing legitimacy to institutions across Europe is strong. This chapter is divided into the following sections: the Bologna process within a Europe of Knowledge strategy; the role and legitimacy of private higher education; the denying of private sector legitimacy and the Bologna denigration of market forces in higher education; and conclusions.

The Bologna Process within a Europe of Knowledge Strategy

Recent attempts at the revitalization of the so-called Lisbon strategy of the European Union (through such widely debated documents like the Wim Kok Report, EC 2004a) seem to be going hand in hand with recent reformulations of the Bologna process in European higher education (Reichert and Tauch 2005). The Lisbon strategy of 2000 is a comprehensive program for increasing EU competitiveness to be implemented in three large areas: economic, social, and environmental. It has set a strategic goal over the next decade: "to become the most competitive and dynamic knowledge-based economy in the world, capable of sustainable economic growth with more and better jobs and greater social cohesion." This goal requires setting up programs for building knowledge infrastructure, enhancing innovation and economic reform, and—of most interest to us here—"modernizing social welfare and education systems" (Lisbon Council 2000, p. 1).[1] The future of Europe seems to be located in a Europe of Knowledge, to be achieved through redefined higher education gained from reformed educational institutions and through boosted research and development in both public and private sectors. New modes of viewing educational institutions are probed (universities as entrepreneurial providers of skilled workforce for the globalizing economy and students as individual clients/buyers of conveniently rendered educational services) and new ideas about citizens gaining enhanced European identity through education useful for knowledge-based Europe are presented (EC 1997; EC 2000d; EC 2003b).

Consequently, in recent years, the project of the European integration seems to have found a new leading motif: education and research for the Europe of Knowledge. At the same time, the Bologna process has been part and parcel of these wider processes of European integration intended to lead to the emergence of the Europe of Knowledge and to the preservation of a distinctive European social model. A crucial component of the Europeanization process today is its attempt to make Europe a knowledge society (and, perhaps even more, a knowledge economy) in a globalizing world. Education and training (to use a more general EU terminology) became a core group of technologies to be used for the creation of a better integrated Europe. The EU has set itself the goal of creation of a distinctive and separate European Higher Education Area and a European Research (and Innovation) Area by the year 2010. The construction of a distinctive European educational policy space—and the introduction of the requisite European educational and research polices—has become an integral part of the EU revitalization within the wide cultural, political, and economic Europeanization project. As Martin Lawn writes, the emergence of a European education area is "fundamental to the contemporary structuring of the EU; it announces the arrival of a major discursive space, centered on education in which the legitimation, steering and shaping of European governance is being played out" (Lawn 2003, pp. 325–326).

We are witnessing the emergence of a new Europe whose foundations are being constructed around such notions as knowledge, innovation, research, education and training. Education, and especially lifelong learning, becomes a new discursive space where European dreams of common citizenship are currently located. However, this new knowledge-based Europe is becoming increasingly *individualized*; ideally, it consists of individual European *learners* rather than *citizens* of particular European nation-states. The emergent European educational space is unprecedented in its vision, ambitions, and capacity to influence national educational policies far beyond the current 25 EU member states. In the new knowledge economy, education policy, and especially higher education policy, it is argued, cannot remain solely at the level of Member States because a new sense of European identity can be best forged only through the construction of a new educational space in Europe. "Europeans," in this context, could refer directly to European (lifelong) learners, individuals investing their dreams for the future in a specific kind of knowledge—knowledge for the knowledge economy.

Clearly then, the Bologna process needs to be viewed in a wider context provided by the idea of the Europe of Knowledge, to be achieved through the implementation of the Lisbon strategy. Most generally, the success of the Bologna process depends on the extent to which it is going to contribute toward the goals of the Lisbon strategy. Its goals, as initially formulated in 2000, were numerous and multidirectional; consequently, most of them were not achievable. Current possible reformulations of the strategy, if it is going to stay alive in the years to come, may include leaving aside both its environmental concerns and most of its social and welfare concerns. The major goals of the strategy would most likely be economic goals, mostly in the spirit already guiding the Lisbon strategy, if not exactly in letter.

Thus the Bologna process is going to be successful if it contributes to the reformulated Lisbon strategy goals, mostly directed toward closer links between education and employability (if not direct employment, as differentiated by Neave 2001) of its graduates, lower unemployment rates, and higher individual entrepreneurship of graduates. Some of the major Bologna goals that today clearly coincide with the goals of the Lisbon strategy include more practically oriented higher education programs, shorter periods of study for the majority of students by a division between the undergraduate and graduate levels, lowering the number of students at the master level, greater intra-European student mobility through various EU-funded mobility schemes, and wider use of credit transfer systems, including within national frameworks.

The Role and Legitimacy of Private Higher Education

The role of the private sector in the countries of Western Europe—where the Bologna process was born—remains marginal. Major EU economies, including Germany, France, Italy, and the United Kingdom, do not have significant private sectors in higher education. But the Bologna process runs far beyond Western Europe to also involve countries where private higher education figures prominently, exceeding 10 percent of total enrollments in Belarus, Bulgaria, Hungary, and Ukraine, 20 percent of enrollments in Latvia, Moldova, Romania, and 30 percent of enrollments in Estonia and Poland.[2] In 2004, over 700 private institutions (including 300 in Poland, 200 in Ukraine, and 70 in Romania) functioned across Central and Eastern Europe, where all countries are already Bologna signatories. In Russia, private enrollments exceeded 14 percent and the number of private institutions reached almost 400. In sum, private sectors present a significant and rapidly developing segment of education—and economy—in Central and Eastern European countries, as testified by Tomusk (2004).

Private institutions in Central and Eastern European countries serve a number of functions, both positive and negative: depending on the country, private institutions may provide fair access to affordable higher education but may also lead to the disintegration of the whole sector, especially if tight licensing and accrediting measures are not in place. These institutions continue to be grappling for legitimacy. The initial social acceptance was strongly impacted by the emergence of many of these institutions in a legal vacuum. Their creation can be attributed both to the enthusiasm for institutional autonomy and the appeal of hitherto nonexistent nonstate educational institutions in new democracies. Currently, most private institutions in the region have been legalized, no longer having to operate on the fringes of the system and are recognized by local national accreditation boards. Their search for social recognition—reflecting the acceptance by the society, the labor market, and their state peers, however, continues.

Private institutions presented the simplest venue toward the expansion of educational systems, which under communist rule were elite. Due to the rapid development of the private sector (and corresponding parallel expansion of the public sector), in some Central and Eastern European countries, higher education became an affordable product. Initial legitimacy of the private sector, in many cases, reflected the social acceptance of the fact that it provided affordable higher education to young people who would not have had a chance to receive it in the closed elite and fully public systems of the former communist countries. In knowledge-based societies, being cut off from affordable education may easily lead to social exclusion and marginalization. Market legitimacy then evaluates the correspondence of the knowledge portfolio received via education with current and future labor market needs. Finally, consumer-granted legitimacy reflects whether services delivered correspond to the personal and professional needs of graduates.

There are Central and Eastern European countries in which the advantages and disadvantages of the existence of the private sector have to be carefully weighed: they have severe problems with the quality of instruction, shortage of qualified (and especially full-time) staff, appearing and disappearing institutions, institutions selling diplomas in a "diploma mill" manner, and so on.

Private institutions are not subsidized by the state except in some cases in some countries; in general, they are almost fully subsidized by students who purchase their teaching services. As a result, the private sector is mostly a teaching sector, with no accompanying research carried out. Consequently, private institutions derive a strong degree of their legitimacy from their students and their families who recognize them as institutions providing services worth paying for. In many cases, being market-driven and consumer-driven in their orientation, private institutions are more flexible to adapt their curricula according to demand, open short-term courses, offer MBA programs, liaise with foreign institutions and offer dual degrees, provide distance education, weekend education, and other modes of learning convenient to the student. Often private institutions monitor the labor market, open career centers for their graduates, and introduce explicit internal quality assurance mechanisms. Many follow market mechanisms in their functioning as business units, use public relations and marketing tools to have significant portions of local, regional, or national educational "markets," and finally prepare their graduates for living and working in market realities. They have also exerted huge impact on academicians themselves.

Many of the above aspects of private institutions in transition countries—and often in contrast to many public institutions—correspond closely to what the Lisbon strategy in general suggests for the education sector in the future. From a certain perspective, it can be argued that most ideas developed in theory in Western Europe and referred to as the Bologna process were actually applied in practice in the private sector in Central and Eastern European countries (those in which the sector exists more than marginally) already in the 1990s, before the ideas of the Bologna process were formulated. The Lisbon strategy in general, and the EU publications about the European Research Area in particular, stress the importance of market forces, individual entrepreneurship of graduates, and new modes of governance of academic institutions; both underlie the perspective of the end-user of knowledge that is the student—rather than its provider, the academic institution. The overall emphasis moves away from the respectable and trustful institution toward the consumer of educational services.

However, the direction of the Bologna process with respect to the Lisbon strategy remains unclear. Above all, the Bologna process seems to downplay the role of market forces in higher education and research and development and omit to consider the private sector that is booming in the transition countries in its overall conceptual scheme.

The Denying of Private Sector Legitimacy and the Bologna Denigration of Market Forces in Higher Education

Despite its intergovernmental (rather than EU) origins, the Bologna process has come to be viewed as an instrument for wider processes of European integration and for wider attempts to preserve a European social model. It is not accidental that there is a common deadline for the Bologna process, the EU Lisbon Agenda of transformations of education, welfare, and economy and the Brugges-Copenhagen process for the integration of European vocational education. The differences between higher education and research in the old EU Member States (EU-15) and the new EU entrants, not to mention other East European Bologna signatory countries, in general, are critical. Higher education in the majority of Bologna-signatory transition countries has been in a state of crisis for over a decade now. While higher education systems in Western European countries seem to face new challenges brought about by the emergence of the knowledge-based economy, globalization, and market-related pressures, most of the Bologna signatory transition countries face old challenges as well, in varying degrees with the need of expansion of their systems at the forefront.

The Bologna process in general seems to focus mostly on new challenges and new problems; most transition countries, by contrast, are still embedded in old-type problems generated predominantly in the recent decade by the need for massification of higher education under severe resource constraints. Bologna-related reforms undertaken in Western Europe are much more functional (fine-tuning, slight changes, etc.); reforms in most Central and Eastern European countries, by contrast, need to be much more substantial (or structural). There is little common ground between the two sets of reforms except for technical details and the Bologna process in its official documents so far seems not to have drawn a sufficiently clear distinction between functional and structural reforms needed in different parts of Europe. Even though the passage to mass systems in Western Europe has been well documented, the current process toward massification of higher education in the transition countries is taking place under different conditions. Therefore, few available recommendations based on the expansion experiences of the EU countries from two-three decades ago exist to the countries of transition. Major suggestions for Western European institutions of higher education may not be sufficient to guide institutions in transition countries. Blind acceptance of the Bologna process and especially blind acceptance of its general conceptual framework may have far reaching consequences for educational systems in these countries. The future of the private sector in the countries of Central and Eastern Europe, despite its controversial role in some of them, is a good example here.

The growing demand for higher education—clearly Daniel C. Levy's "demand-absorbing" wave of the growth of the private sector—gave rise to the booming of private higher education institutions in several transition countries. While, apparently, the rapidly developing private sector seems of marginal importance for the Bologna process in Western Europe (and perhaps therefore it has not been dealt with in the Bologna documents so far), it certainly is a problem (and/or solution) for some transition countries, where private sectors play a significant role, including Poland, Ukraine, Estonia, and Romania. The rapid development of private higher education as well as the emergence of powerful market forces on the educational and research landscape of most transition countries, I believe, require further analyses, and, consequently, the consideration into the debates accompanying the implementation of the Bologna process on a European scale. So far, by ignoring the booming private sector in the countries of Central and Eastern Europe, and thereby ignoring powerful market forces and market mechanisms in higher education there, the Bologna process appears to be indirectly refusing legitimacy to institutions of private higher education. The fact that the Bologna process does not use the word "market" or the word "private," in transition countries that still have their systems in flux, and have no guidance on how to expand access to higher education under severe state underfunding, suggests a refusal to grant legitimacy to the sector, an indirect rejection of the competition between the public and private sectors, and an implicit suggestion that the existence of market mechanisms in teaching and research is fundamentally wrong.[3]

Yet, it is the private institutions, which, especially in transition countries with larger private sectors, are often closer to the recommendations of the Lisbon strategy than public institutions. And the Bologna process becomes increasingly part of a much larger social and economic transformation of Europe epitomized by this strategy. The Bologna game in higher education is the most powerful game in town in most transition countries; for most governments in these countries, it provides the best rationale available for reforming the systems. The number of signatory countries already exceeds 40. Bologna provides a major impetus for the otherwise often static systems, and the idea of catching up with a larger European trend is often much better received by the general public in these countries than in Western European countries. More so, in some non-EU transition countries following Bologna requirements is even regarded as bringing the country closer to the EU, or seen as a temporary substitute for EU membership.

As a result, the Bologna process is one of the most ambitious transformations of higher education systems on a regional scale in the world today. Its impact on the future of European higher education is potentially deep and long-lasting (as is potentially the impact of the emergence of the European Research Area, discussed in Kwiek 2006). Since the very beginning, the Sorbonne Declaration, through the Bologna Declaration—Prague and Berlin summits (2001 and 2003), as well as from the Salamanca (2001) to the Graz (2003) to the most recent Glasgow (2005) declarations of higher education institutions, the private sector has been neglected as a topic of educational analysis. As an example, in the most recent European University Association's Glasgow Declaration "Strong Universities for Strong Europe," the word "private" appears once (private funding), and the word "market" appears twice: labor market and employment market (*Glasgow Declaration* 2005). For the official documents and accompanying reports of the Bologna process, the private sector does not exist. While declarations and communiqués of the Bologna process have not made a single reference to private higher education in the last seven years, *Trends III* report of 150 pages (prepared for the Berlin summit of Bologna signatories in 2003) mentions the term half-a-dozen times but only in connection with the GATS negotiations, as if the issue of the rapidly emerging private sector and an increasing market orientation of higher education institutions both globally and in many Bologna signatory countries were irrelevant. The situation is not different in the recently published *Trends* IV report: the word "private" appears four times but never in connection with the higher education sector; the word "market" appears more than a dozen times but almost exclusively in relation to the labor market. There are also no indications that the notion of "competition" is taken seriously by the report, either in its spirit or as part of the vocabulary used in the body of the text (Reichert and Tauch 2005). The omissions go against global trends in which the role of the private sector in teaching and research is on the rise, market forces are a significant part of the educational and research landscape, and the competition for students, public, and private research funds, and competition between institutions and faculty is an important factor (Altbach 1999; Levy 2003).

Consequently, the ideas behind Bologna, the analytical tools it provides, the wider picture of the role of higher education in society and economy, and policy recommendations it develops may have unanticipated and mixed effects on higher education systems, especially in Eastern (rather than Central) Europe where it is still possible to grant or refuse legal legitimacy, for example through new legislation. To be an effective integrating tool on an European scale, the Bologna process would need to take into account the fundamental difference between Western European countries and some transition countries with respect to the role of the private sector and the role of market mechanisms in higher education. In most transition countries (especially in Central Europe), private institutions currently play a significant role.

At the same time the role of the private sector in the countries of Central and Eastern Europe—considering its ability to adapt to the new societal needs and new market conditions combined with the drastically underfunded and still unreformed public institutions with limited capacities to enroll larger numbers of students—and despite its lack of recognition on the part of the Bologna process is bound to grow. Private institutions represent a wide variety of missions, organizational frameworks, legal status, and relations to the established institutional order. What is needed is the disinterested analysis of the current (in-transition) state of affairs, largely unexplored so far in international educational research, and conclusions as to how to deal, in theory and in practice, with growing market forces in education; how to regulate privatization and corporatization of educational institutions and research activities within ongoing reform attempts, and finally how to accommodate principles of the European Research Area and requirements of the Bologna process to local conditions of new EU countries. Unfortunately, the Bologna process in its current form, in general, remains indifferent to these developments even though their appearance in transition countries might prefigure many future options that Western European policymakers might face.

Conclusions

The refusal of legitimacy to private higher education and the market forces in education in general within the Bologna process may lead to a limitation in the expansion of higher education system as a whole, in numerous Central and Eastern European countries where the private sector has not been developed so far. In Central and East European transition countries, educational business is increasingly private, teaching-focused, and market-driven. It does not seem to change the substance of the implementation of the Bologna process but it does affect the overall functioning of the two sectors in transition countries and consequently the effectiveness of Bologna reform strategies. There is a strong market-driven competition for students between private institutions, and a strong competition for faculty between private and public institutions. Transition countries, generally, have to start or already cope with the rapid massification of their systems, with the number of students being on the rise. The Bologna process has been developed for Western European countries and now it is being implemented from Portugal to the Caucasus. In most of them, the process is viewed in terms of "catching up" with the West, quite often as a substitute for the political integration. At the same time, long-term consequences of the process for national education systems with vastly different problems are unclear. Unfortunately, major Bologna-related documents do not seem to take the problem of both the private sector and the market forces in higher education into account. The overall revitalization of the European integration project through education, and the accompanying production of the new European citizenship, may bring about unexpected effects in transition countries in which welfare state regimes are different, higher education systems and labor markets have their own traditions, and which generally are at different stages of economic development. Strong private sector and powerful market forces can be viewed as good examples of significant (but so far neglected) differences between the countries where the Bologna ideas were born and the countries where these ideas are currently, almost unanimously, implemented.

Private higher education and strong market forces in education in transition countries require careful analysis in European educational research. Little known in the old EU-15 (except for example Portugal, the Netherlands, or the United Kingdom), they may indicate more global trends and tendencies, to be seen in the old EU-15 in the future. Both serious problems and excellent solutions brought about by the private sector in transition countries deserve careful research attention. The Bologna process, neglecting these developments, is an example of how experiences in the peripheral European countries can be out of research focus today.

Notes

1. The shift to a digital, knowledge-based economy is a powerful engine for growth and competitiveness, the strategy argues. Consequently, the document affirmed the idea of a European Area of Research and Innovation. The necessary steps mentioned for the education sector include e.g. developing mechanisms for networking, improving the environment for private research investment, research and development partnerships and high technology start-ups, encouraging the development of an "open method of coordination" for the benchmarking of national research and development policies, taking steps to increase the mobility of researchers and introducing Community-wide patents (Lisbon Council 2000, pp. 3–4). The targets set in Lisbon for education included a substantial annual increase in per capita investment in human resources, the number of 18 to 24 year olds who are not in further education and training to be halved by 2010, schools and training centers to be developed into "multi-purpose local learning centers" accessible to all, and the development of a European framework defining the new basic skills to be provided by lifelong learning and defining a common European format for curricula vitae.
2. Consider, for example, the robust growth of the Polish private higher education sector. Until the collapse of Communism in Poland in 1989, higher education there was fully controlled by the state. A new Higher Education Act of 1990 paved the way to the development of the private sector in general and a Vocational Higher Education Schools Act of 1997 provided legal grounds for lower-level vocational private sector. The number of private institutions rose from 3 in 1991 to 250 in 2002 and 301 in 2005. Since the beginning of the 1990s, the private sector has changed the educational landscape in Poland beyond recognition. In the last decade and a half, the number of students rose more than four times, from about 400 000 in 1990/1991 to over 1 900 000 in 2004/2005. In academic 2004/2005, almost one third of the student body (over 30 percent) went for private higher education institutions. In recent years, private higher education institutions have been developing smoothly and under the close supervision of the Ministry of Education. They have become a challenge to public institutions. Their increasing number has also increased the accessibility of the higher education system as a whole. Private institutions, especially in smaller towns, often provide the only available form of higher education (which is also cheaper than public education in university cities when accommodation costs are taken into account).
3. The strangeness of omitting private dynamics is illustrated by data on public funding for higher education. Poland's public funding (1995–2004) has generally been between 0.8 and 0.9 percent of GDP, a figure slightly lower than those in other EU countries (For 2001, from 0.8 in Italy and the United Kingdom up to 1.5 in Sweden and 1.8 in Denmark, respectively (combined with private funding, the percentage of GDP for education in these countries was: 0.9 in Italy, 1.0 in Germany, 1.1 in France and the United Kingdom, 1.2 in Spain, 1.3 in the Netherlands and Ireland and 1.8 in Denmark). The highest percentage of private funds spent on higher education as a share of GDP has been 0.3 (Spain, Ireland, and the United Kingdom).

Bibliography

Altbach, Philip G. (ed.). 1999. *Private Prometheus: Private Higher Education and Development in the 21st Century*. Chestnut Hill: CIHE.

Bologna Declaration. 1999. Joint Declaration of the European Ministers of Education Convened in Bologna, June 19 in Council of Europe 2002.

Council of Europe. 2002. Compendium of Basic Documents in the Bologna Process. Strasbourg: Council of Europe.

European Commission. 1993. Growth, Competitiveness, Employment. The Challenges and Ways Forward into the 21st Century. White Paper. Brussels. COM93700.

——— 1997. Towards a Europe of Knowledge. Brussels: COM97563.

——— 2000d. A Memorandum on Lifelong Learning. Commission Staff Working Paper. Brussels. SEC2000 1832.

——— 2003a. Investing Efficiently in Education and Training: An Imperative for Europe. Brussels. COM 2002779.

——— 2003b. The Role of Universities in the Europe of Knowledge. Brussels. COM 200358.

——— 2004a. Facing the Challenge. The Lisbon Strategy for Growth and Employment. Report from the High Level Group, chaired by Wim Kok. Luxembourg: Office for Official Publications for the EC.

——— 2004b. Achieving the Lisbon Goal: The Contribution of VET. Final Report to the European Commission. Brussels. November 1, 2004.

Graz Declaration. 2003. *Graz Declaration. Forward form Berlin: The Role of Universities*. Brussels: EUA.

Graz Reader. 2003. EUA Convention of European Higher Education Institutions. Strengthening the Role of Institutions. Brussels: European University Association.

Kwiek, Marek. 2001. Social and Cultural Dimensions of the Transformation of Higher Education in Central and Eastern Europe. *Higher Education in Europe*, Vol. 26, No. 3. 399–410.

——— (ed.) 2003a. *The University, Globalization, Central Europe*. Frankfurt a/Main-New York: Peter Lang.

——— 2003b. Academe in Transition: Transformations in the Polish Academic Profession. *Higher Education*, Vol. 45, No. 4, June. 455–476.

——— 2004a. *Intellectuals, Power, and Knowledge. Studies in the Philosophy of Culture and Education*. Frankfurt a/Main and New York: Peter Lang.

——— 2004b. The Emergent European Educational Policies Under Scrutiny. The Bologna Process From a Central European Perspective. *European Educational Research Journal*, Vol. 3. No. 4, December. 759–776.

——— 2004c. The International Attractiveness of the Academic Profession in Europe. Country Report Poland. In Jürgen Enders and Egbert de Weert (eds.), *The International Attractiveness of the Academic Workplace in Europe*. Frankfurt a/Main: GEW. 332–349.

——— 2005. Renegotiating the Traditional Social Contract? The University and the State in a Global Age. *European Educational Research Journal*, Vol. 4, No. 4., December. 324–341.

——— 2006. *The University and the State. A Study into Global Transformations*. Frankfurt a/Main and New York: Peter Lang.

Lawn, Martin. 2001. Borderless Education: Imagining a European Education Space in a Time of Brands and Networks. *Discourse: Studies in the Cultural Politics of Education*. Vol. 22, No. 2. 173–184.

——— 2003. The "Usefulness" of Learning: the Struggle over Governance, Meaning and the European Education Space. *Discourse: Studies in the Cultural Politics of Education*, Vol. 24, No. 3. 325–336.

Levy, Daniel C. 1986. *Higher Education and the State in Latin America. Private Challenges to Public Dominance*. Chicago: The University of Chicago Press.

——— 2002. Profits and Practicality: How South Africa Epitomizes the Global Surge in Commercial Private Higher Education. PROPHE Program for Research on Private Higher Education Working Papers Series. No. 2. at www.albany.edu/~prophe http://www.albany.edu/dept/eaps/prophe/publication/paper/PROPHEWP02_files/PROPHEWP02.pdf, retrieved on January 10, 2006.

——— 2003. Unanticipated Development: Perspectives on Private Higher Education's Emerging Roles. PROPHE Program for Research on Private Higher Education Working Papers Series. No. 1 at www.albany.edu/~prophe http://www.albany.edu/dept/eaps/prophe/publication/paper/PROPHEWP01_files/PROPHEWP01.pdf, retrieved on January 10, 2006.

Lisbon Council. 2000. Presidency Conclusions. Lisbon European Council 23 and 24 March 2000 at http://ue.eu.int/ueDocs/cms_Data/docs/pressData/en/ec/00100-r1.en0.htm, retrieved on January 10, 2006.

Main Statistical Office (GUS). 2005. *Higher Education and Its Finances in 2004*. Warsaw: Main Statistical Office Press (in Polish).

Neave, Guy. 2001. Anything Goes: Or, How the Accommodation of Europe's Universities to European Integration Integrates—An Inspiring Number of Contradictions. Paper delivered at EAIR Forum, Porto, September 2001.

Radó, Péter. 2001. Transition *in Education. Policy Making and the Key Education Policy Areas in the Central European and Baltic Countries*. Budapest: OSI.

Reichert, Sybille and Christian Tauch. 2003. *Trends in Learning Structures in European Higher Education III*. European University Association. http://www.eua.be/eua/jsp/en/upload/Trends2003final.1065011164859.pdf, retrieved on January 10, 2006.

——— 2005. Trends IV: European Universities Implementing Bologna. European University Association. http://www.eua.be/eua/jsp/en/upload/TrendsIV_FINAL.1117012084971.pdf, retrieved on January 10, 2006.

Tomusk, Voldemar. 2004. The Open World and Closed Societies. Essays on Higher Education Policies "in Transition." New York: Palgrave.

UNESCO. 2005. Statistical Tables. Gross Enrolment Ratios, Tertiary at http://www.uis.unesco.org/TEMPLATE/html/HTMLTables/education/ger_tertiary.htm, retrieved on January 10, 2006.

UNESCO-CEPES. 2004. Statistical Information on Higher Education in Central and Eastern Europe 2003–2004 at http://www.cepes.ro/information_services/statistics.htm, retrieved on January 10, 2006.

Part II
Country Perspectives

Chapter Six

Legitimacy Discourse and Mission Statements of Private Higher Education Institutions in Romania

Robert D. Reisz

Introduction

Legitimacy, an important and well-studied concept in organizational sociology, has been approached from different perspectives. Probably the most widely accepted definition of legitimacy stems from Suchman. According to him "legitimacy is a generalized perception or assumption that the actions of an entity are desirable, proper or appropriate within some socially constructed system of norms, values, beliefs and definitions" (Suchman 1995, p. 574).

While this general understanding of legitimacy can be traced back to Weber's classification of legitimate authority, a recent summary of the literature on organizational legitimacy (Suddaby 2002) finds that research has concentrated upon three aspects to organizational legitimacy that only partly parallel Weber's three types:

- "an important *regulatory/legal* component, in which the legitimacy of a particular organizational form depends upon its conformity with explicit rules and regulations;"
- "a strong *moral* element to legitimacy in which new organizational forms or practices are evaluated against commonly held values and beliefs and shared assumptions about whether the action or structure is 'good' or 'bad';"
- "*economic* prowess or *technical* efficiency is an important determinant of the acceptability of an innovation" (Suddaby 2002, pp. 2–3).

Another important aspect noted in an overview of legitimacy by Slantcheva (2004) is that of "the sources of legitimacy—or which groups or institutions have the authority to confer their approval on an organization or its practices of a given type." The three elements of legitimacy mentioned by Suddaby have a clear societal aspect. Those empowered to grant legitimacy include official actors in the regulatory/legal component of efficiency as well as societal opinion-formers. While economic prowess is of lesser importance to the subject of this study's interest, the moral element of legitimacy becomes central to the concerns of higher education institutions.

Legitimacy bolstering is particularly important for new institutions that strife/aspire to be accepted as what they claim to be. The survival of new organizational forms, as is the case with the new private higher education institutions, is critically dependent on their successful legitimation (Hannan and Freeman 1989). Private higher education institutions have to address both the legal elements to legitimacy, represented first and foremost by the accreditation procedures, and the moral elements. Private institutions in Romania and elsewhere (e.g., in Central and Eastern Europe, the former Soviet states, China, South Africa, India) have not been readily accepted by society as part of the university sector. This development has been a result of the prevailing traditional concept of a university as a place where knowledge is created—above all—disseminated and distributed (Pellert 1999). The new private institutions have not been seen as places where knowledge is created. Moreover their reliance on tuition fees in order to perform their function of knowledge distribution has been often considered morally questionable.

Thus, on the one hand, the legitimation efforts of the emerging private sectors in Central and Eastern Europe have been largely in response to their novelty and their privateness (Levy 2005). On the other hand, legitimacy is vital for these institutions. The services they market need the recognition both of their outcomes in the form of degrees and diplomas and of status allocation. Legitimacy comes forth as a pseudo-currency. In a groundbreaking article, John W. Meyer's views "the institutional effects of education as a legitimation system" (Meyer 1977, p. 55). Meyer develops Bowles and Gintis' idea (1976) according to which "education is thought to socialize people to accept as legitimate the limited roles to which they are allocated (Meyer 1977, p. 64). This concept of an educational system recognizes that the society at large has to accept the outcomes of the educational system as legitimate. If education is a system that legitimates individuals into social positions and thus makes them also act according to their pre-determined status, the educational institutions that empower the individual need to have a form of legitimacy themselves. Put in the metaphorical terms of symbolic economy, the legitimacy held by the institution is sold for a fee to the student, to legitimate that student as a higher education graduate. If the institution has questionable legitimacy, so will its graduates.

In this context, new private higher education institutions develop strategies to gain legitimacy in order to survive. These strategies materialize in discourse and action that fluctuate between mimetic isomorphism (DiMaggio and Powell 1991) with legitimate institutions and attempts to justify their existence through specialization. In other words, on the one hand, private institutions try to copy legitimate institutions, public or older private ones in order to "prove" they are indeed higher education institutions. On the other hand, if private institutions would cease to be different from public ones or new institutions from the older ones, their *raison d'être* would cease to exist (Whitehead 1977).

Suddaby (2002) finds that the major efforts in understanding legitimacy focus on linguistic data. Discourse is evidently a means by which the social world is constructed (Berger and Luckman 1966). Organizations as well as individuals develop discourse to describe, evaluate, explain, and diagnose reality. The institutionalization of such discourse patterns in organizations shapes the organizations as well as the way they are perceived (Manning 1992). Linguistic data will also be the most important analytical source to this chapter.

In the following sections, I consider the mission statements of Romanian private higher education institutions as part of their efforts to gain legitimacy. I view them as elements of discourse that both describe and shape reality. Mission statements are short summaries and, as such, very suitable for discourse analysis. They constitute an element in legitimacy construction from at least two points of view. On the one hand, they intend to legitimate the organization through their content, proving that the organization has a legitimate mission to fulfill, a reason to exist. On the other hand, the mission statements are part of the legitimating effort through their form and wording.

Methodology

This chapter is based on quantitative and qualitative analysis of the Web pages of Romanian private higher education institutions. Web pages have been chosen as the empirical basis due to their role of carriers of institutional discourse. They are a medium for advertisement as well as an opportunity for the systematic self-presentation of institutions. The information on the web pages not only represents the organizational effort for self presentation, but is also indicative of the information selection preferences. It gives access to the way private higher education institutions see themselves and want to be seen by others. In addition, one can find what information the institution considers important to advertise.

The list of institutions I used originates from the National Council of Academic Evaluation and Accreditation (NCACEA) in Romania from January 2004. This independent buffer body gives temporary authorization, and then proposes for accreditation, higher education institutions in Romania. The list includes 23 accredited institutions and 28 authorized institutions.[1]

At the time of the study, of all 51 institutions, 38 had Web pages. The remaining 13 institutions either had no pages at all or had addresses that were not accessible on repeated occasions during a month period. The pages provided

- quantitative data presented by the institutions themselves, including the number of faculties, number of students and teaching staff, and year of creation of the institutions. In some cases, information was filled in from sources of the NCAEA and the CALISRO (Quality in Romanian Higher Education) program;
- the mission statements of the institutions. Not all institutions had explicit mission statements. In some cases, I considered as mission statements fragments from welcome speeches of the rectors or founders, or short presentations included under titles like "who are we," "about us" or other similar self-descriptions that had the content of mission statements.

This chapter is based on this collected data. It provides information on the developments of the private higher education sector in Romania and, on the basis of the results from the empirical study, attempts to analyze the mission statements.

Development and Basic Characteristics of the Romanian Private Sector

The Romanian higher education system is rather young compared to those of other Central European countries. Though the first university, the Jesuit University in Cluj, was created at the end of the sixteenth century, the modern higher education system finds its roots in institutions founded at the beginning of the nineteenth century. According to Scott (2000), Romania was the only country in Central and Eastern Europe whose institutions of higher education had been inspired both by the Napoleonian and the Humboldtian models. He alludes to the Mihailian Academy in Iasi (founded in 1830) and the High Technical School in Bucharest (founded in 1818). In the years after 1859, when the unification of two of the Romanian principalities took place, the university system was reformed and the current institutions emerged.

The higher education system underwent an important expansion in the 1930s and, still an elite system, entered the post-war reforms undertaken by the Communist Party similar to the rest of Central and Eastern Europe. During the post-war years, the system expanded and suffered the politechnization imposed by the communist view of modernization and transformation of higher education in the production of intellectual proletariat. The last two decades of the communist period included elements of increased rigidity. Curricula remained unchanged and a recession in student numbers took place while no new staff was hired (Reisz 2003a).

As a result, in 1990 Romania had a much underdeveloped higher education system occupying the second but last position in student numbers per 10,000 inhabitants in Europe (Ladanyi 1991). This accumulated lag in inclusion in the higher education system led to the explosive developments after 1990 and to the creation of a private sector (Sadlak 1994). In the beginning, the private sector mushroomed in a legislative vacuum (Reisz 1992). A law for the accreditation of higher education institutions was passed in 1993, followed by an education law in 1995 (Reisz 2003a). These legal acts led to a certain stabilization of the uncontrolled developments of the system. In the following years, the public sector regained its institutional autonomy. Study fees introduced in the public sector as well as the accreditation procedures made the public and the private sectors converge to a certain extent. This convergence additionally contributed to the strategy of many private institutions of copying public institutions (Dima 1998). Isomorphism, resulting from the copying of successful or legitimate models (DiMaggio and Powell 1991) of public higher education institutions by private ones, has, in fact, been previously analyzed and found to be a major reason for the lack of institutional diversification brought by the emergence of private sectors (Levy 1999).

At the beginning of the twenty-first century, roughly one-third of the student population was enrolled in the private sector in Romania. Yet, most private institutions had lower prestige than any of the public ones and a more fragile market position.

With respect to legitimacy, three indicators need to be considered in the Romanian private higher education sector: age of institution, size, and disciplinary structure. The first two cover issues of organizational fragility, newness, and smallness that have been theoretically developed in organizational ecology, and the last relates to the elements of higher education mission.

1. Age is directly related to organizational legitimacy, as new institutions are particularly fragile ("the liability of newness:" Hannan and Freeman 1989). Even more so, new higher education institutions have to face the socially constructed image of the relationship between academia and tradition that is particularly important in Central Europe.

2. Size is less related to legitimacy in the academic world, but is one of the elements of organizational fragility in general ("the liability of smallness": Hannan and Freeman 1989).
3. Disciplinary structure: as Parsons (1956) notes, legitimacy results from a relationship between organization and its social environment. Thus, the legitimacy of higher education institutions results from their confrontation with the social definition of higher education. The disciplinary span of private higher education, centered on vocational subjects, relates to this confrontation.

The years of foundation of Romanian private higher education institutions reveal the steep organizational expansion of private higher education between 1990 and 1992. During this period 33 private institutions were created, but having not received accreditation subsequently, lost their temporary license and ceased to exist. There existed three confessional institutions predating 1990. After 1992, the number of newly established private institutions never again reached the peaks from the beginning of the decade. Similar patterns in the development of the private higher education sectors in Central and Eastern Europe in the 1990s have also been mentioned by Levy (2004).

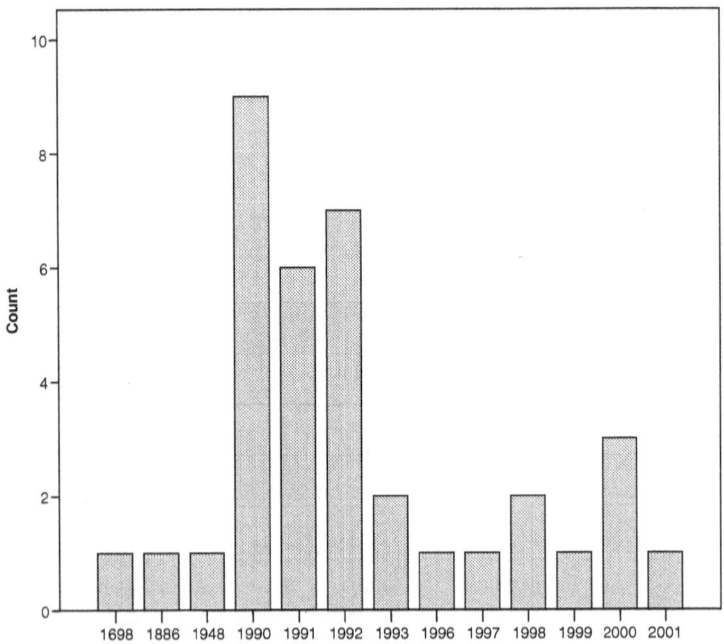

Establishment of private higher education institutions, by year of foundation

A small number of the existing private institutions have more than 4 faculties (table 6.1) while the remaining large majority of institutions have fewer faculties than 4. The institutions that have a larger number of faculties (more than 10) are all multi-campus institutions having usually faculties in the same disciplines on different campuses. This feature is typical of private higher education institutions internationally, where private institutions are often smaller in size and with narrower disciplinary span than the public universities. While the number of faculties is not directly dependent on either size or disciplinary span of higher education institution, it correlates highly with them.

Only 22 institutions provided information on the number of students on their Web pages. It is probable that most institutions that have not published such information are closer to the lower end of the scale. Compared to public institutions, size of both student enrollments and faculty in the private higher education institutions is smaller (table 6.2 and table 6.3).

The information on disciplines taught in private higher education institutions (table 6.4) comes from the CALISRO program for quality in higher education. The data include the disciplines of all authorized study programs without listing them under the faculties they were offered at (faculties can include study programs in different categories). The rough data collected from CALISRO had a very large number of disciplines. I had to group these to get to the manageable categorization presented in table 6.4. The categories prove

Table 6.1 Number of private institutions according to number of faculties

Number of faculties in the institution	*Number of institutions*	*Percentage*
1	7	13.7
2	8	15.7
3	14	27.5
4	7	13.7
5	4	7.8
6	1	2.0
7	3	5.9
8	1	2.0
9	1	2.0
10	1	2.0
14	1	2.0
18	1	2.0
21	1	2.0
28	1	2.0
Total	51	100.0

Note: In the statistical tables percentages do not add up to 100 due to rounding.

Source: Data collected according to methodology (p. 137).

Table 6.2 Number of private institutions according to number of students

Number of students	*Number of institutions*	*Percentage*	*Cumulative percentage*
Under 300	4	7.8	18.2
301–1000	3	5.9	31.8
1001–5000	8	15.7	68.2
5001–10000	4	7.8	86.4
Over 10000	3	5.9	100.0
Total	22	43.1	
Missing Data	29	56.9	
Total	51	100.0	

Source: Data collected according to methodology (p. 137).

Table 6.3 Number of private institutions according to number of teaching staff

Number of teaching staff	*Number of institutions*	*Percentage*	*Cumulative percentage*
Under 100	10	19.6	62.5
101–300	4	7.8	87.5
Over 301	2	3.9	100.0
Total	16	31.4	
Missing Data	35	68.6	
Total	51	100.0	

Source: Data collected according to methodology (p. 137).

Table 6.4 Categories of study programs in private higher education institutions

	Number of study programs	*Percentage*
Economics and business studies	99	39.1
Law	23	9.1
Social sciences	17	6.7
Teology	16	6.3
Languages	14	5.5
Journalism	12	4.7
Medicine	11	4.3
Information S & T	10	4.0
Political science	10	4.0
Arts	9	3.6
Sports	9	3.6
Humanities	7	2.8
Engineering	7	2.8
Natural Sciences	7	2.8
Teacher Training	1	0.4
Architecture	1	0.4
Total	253	100.0

Source: Data collected according to methodology (p. 137).

to cover well the disciplinary structures represented in private higher education both in the Central and Eastern European region and also worldwide with their focus on economic and business studies, law, and social sciences (Levy 1992).

It is important to mention that the economics and business studies category does not include a program named economic sciences. The most frequent study programs are in finances and banking (19), management (18) and accounting (17). The overwhelming majority of study programs have a professional/vocational orientation. The majority of science programs are in the social and political sciences, disciplines that did not exist in Romanian higher education prior to 1990, being in very high demand after the changes.

Legitimacy Discourse and Mission Statements

Categories of Legitimacy Discourse and Organizational Traits

A way to handle the different mission statements of Romanian private higher education institutions was through categorizing them and analyzing the particularities of the different categories. I have developed the categories according to mission statements, other information from the Web pages, and names of institutions, all elements of the legitimacy discourse. In this section, I analyze organizational traits of the institutions as well as the variance of mission statements within each category.

In the private higher education literature, studying the variance in institutional character through categorization has been used before (Geiger 1986; Levy 1986, 2005). I have previously classified the legitimation discourse of Central and Eastern European private higher education institutions in six empirical real types (Reisz 2003b). Compared to the method of ideal types, real types, also defined by Max Weber, have existing counterparts or are supposed to exist themselves. The classification of institutions around these real types has to be regarded more as a form of clustering than as belonging. The categories are

- Confessional higher education: In addition to the old confessional higher education institutions, new institutions formally related to the church as well as institutions that present a religious mission without formally being related to a religious organization exist.
- "International"[2] higher education institutions: include in their names "international," "European" or clearer references like "American," "German," and so on. These institutions seek to legitimize their programs, mainly in economics, business and law in relation to external, mostly American models of business and law schools.
- Regional higher education institutions are legitimated by their relation to local-regional demands.

- Diversity institutions are institutions that claim legitimacy for being different. These institutions often base their mission on ideologies, such as environmentalism, Waldorf education, and so on.
- Institutions owned by enterprises have generally tried to gain legitimacy from the image of the founding institution and sustain missions related to the profile of the founder.
- Public-isomorphous institutions are institutions that argue to be no different from public universities (Reisz 2003b, pp. 29–30).

One goal of the present research was to also test these previously established categories by finding their distribution in the population of Romanian private higher education institutions. These categories were intended to discriminate among institutions according to their legitimacy discourse expressed through their mission statements, the names of institutions, and the overall outline of their Web pages, where these were available. In some cases, I had to contend with the names and short presentations available at the Accreditation Committee and the CALISRO program. Nevertheless, the categories are of legitimacy discourse. My comments deal with the organizational characteristics of institutions that have a particular legitimacy discourse. I also comment, in some cases, on the forms that mission statements take within a certain category of legitimacy discourse to justify the categorization itself and show the existing variance within groups (see table 6.5).

Confessional private higher education has been of high importance in all traditional private higher education sectors, from Latin America to Belgium or Hungary; still, it is relatively surprising, how high an importance, confessional institutions have gained in the Romanian private sector in recent years. Of the 11 institutions, 3 are traditional institutions founded before the communist rule, but 8 are newly created institutions. Most confessional institutions offer only theological education, pastoral theology, as well as

Table 6.5 Number of institutions according to institutional legitimacy discourse

Category of legitimacy discourse	*Number of institutions*	*Percentage*
Confessional	11	21.6
International	3	5.9
Regional	18	35.3
Diversity	5	9.8
Enterprise-owned	4	7.8
Public-isomorphous	10	19.6
Total	51	100.0

Source: Data collected according to methodology (p. 137).

the preparation of religion teachers. The religion teachers are also prepared to educate in some other fields of the arts, humanities, or social sciences. All Roman-Catholic institutions as well as the Protestant university that pre-existed the 1990s have a religious disciplinary focus. Three of the confessional universities also include study areas that are not related to religious education, these being economics, humanities, arts, or law. The distribution of institutions within the confessional category is as follows: four institutes are Roman Catholic (amongst which are two of the traditional institutions), four are neo-Protestant (two Baptist, one Adventist, one Pentecostal), and two are Protestant (one traditional), while one institution is Christian ecumenical. It should be mentioned here that, even if the Greek-Orthodox faith accounts for 80 percent of the Romanian population, no private higher education institution represents this confession. One reason for this could be found in the large number of Greek-Orthodox theology faculties in public universities. Roman-Catholic, Protestant, and neo-Protestant faculties also exist in public institutions but their numbers and size are much lower. Confessional private institutions have generally kept a lower profile than the rest of the private sector. This is definitely to be related to their existing market niche, the degree of legitimacy, and, in some cases, a longer institutional history.

There are three "international" institutions that claim their legitimacy from external models. One is American, one British, and one German. The first two are in Bucharest, while the last one is, predictably, in Sibiu (Hermannstadt), a city where the renewal of German Saxon traditions has recently become very important leading to the election of a German mayor, the development of German language education, and the attraction of important German investment both in the economy and in the renovation of the medieval old town. All three "international" institutions import curriculum, host visiting professors from the respective countries, and organize study visits abroad.

Regional private higher education institutions make up the majority of the Romanian private sector. I have 18 institutions included in this category. The legitimating efforts of these institutions relate to local-regional communities in different ways and degrees, in some cases, the inclusion of an institution into this category was not self-evident. Some institutions define themselves as "community higher education institutions" (2 institutions), 4 other institutions specify in their mission-statements their existence as universities "for" the specific region ("a university for the South-East of the country") or preparing specialists for a region ("for the Western part of Romania"). Other regional institutions include in their presentation references to the region of location, or claim priority in the private higher education sector of the region ("the first private university in Moldova"). Some institutions make references to the relevance of their degrees and the education they offer for the regional or local labor market.

Most regional institutions have emerged after the first wave of private higher education institutions in the early 1990s or have developed from branches of institutions founded in the major academic centers. While in the beginning, such institutions were upheld with the help of commuting academics, once the accreditation procedures came into force, the institution's own academic staff had to be created. Most such institutions are located in cities different from the traditional academic centers. Nevertheless, there exist institutions in Timisoara or Iasi that explicitly declare themselves as having a regional focus. Many regional institutions that have been created in cities where no public institutions exist have a relatively good market position as the costs of education at the private institutions are compared with cost of living, travel, and education at public institutions in other cities.

Diversity institutions have become rare after a somewhat higher importance and/or visibility in the early 1990s. The number of institutions that fit in this group are 5. Of these, 2 are "ecological" universities, 1 is a college that prepares teachers for Waldorf schools, 1 is a university for the Hungarian minority, while the last claims to prepare businessmen. This last institution has been included here as its legitimating discourse concentrates on the "difference" to other institutions. These institutions intend to offer something different and their presentations and mission statements are directed toward their being different from other institutions. In this respect, they tend to have a discourse similar to the "international" institutions.

Four Romanian private institutions are owned or created by enterprises and develop their legitimacy from the prestige of their founders. These institutions have quite a different status and a different disciplinary relevance. The founders also embed the institution in the respective profession. These cases are: an institute for banking created by the National Bank and the Association of Romanian Banks, a school of journalism (including television, radio, etc.) owned by one of Romania's largest private media groups, a college for dental assistants and technicians that is related to a large dental clinic, and a college owned by the association of Handicraft Co-operatives of Romania. The last of these 4 institutions is closer to the regional or public-isomorphous institutions, offering typical management and marketing studies. The other 3 institutions are very different from "regular" private higher education, and can be considered on the higher end of the prestige scale of Romanian professional higher education.

Finally, 10 institutions fall in the public-isomorphous group. These institutions are, as a rule, located in Bucharest, have a larger number of students, a larger number of faculties, and a wider range of disciplines. They declare themselves as having the same (or better) standards as the public institutions. Almost all study programs in the sciences are offered by these institutions. I have called these institutions public-isomorphous as their

Table 6.6 Average size in numbers of students according to category of legitimacy discourse

Category	*Average size*	*Number*
Confessional	3119.60	5
International	2763.33	3
Regional	4828.57	7
Diversity	1500.00	1
Enterprise-owned	430.00	2
Public-isomorphous	11750.00	4
All institutions	4865.82	22

Source: Data collected according to methodology (p. 137).

claim for legitimacy is based on their being no different from public universities. All these institutions have been founded in the early 1990s.

A comparison of the average institutional size with respect to numbers of students within the developed categories leads to interesting results even if data could be collected only for part of the institutions. With an average of over 10,000 students, public-isomorphous institutions have dimensions similar to those in the public sector. Enterprise-owned institutions are extremely small, which has also to be related to their disciplinary range. Somewhere in between are the remainder of the institutions, regional institutions tending to be somewhat larger than confessional or "international" ones (see table 6.6).

Content of Mission Statements

In this final section I take a closer look at the collected mission statements. If until now I have analyzed organizational characteristics of institutions with different categories of institutional discourse, I now proceed to analyze the wording and content of the mission statements.

The form of the mission statements, like the design of the Web pages, tries to offer an image of *confidence and reliability*. The rhetoric of the statements is generally formal, correct, and somewhat bureaucratic. There are very few exceptions where some poetic elements and religious rhetoric find their way. None of the statements seems to address prospective students but rather parents or eventually other academics or administrators. The mission statements are filled with words like "community," "professionals," "standards," "specialists," "leaders," "calling," "entrepreneur," and so on, words that are part of the new official rhetoric. The statements are generally dense in neologisms, complying with certain popular expectations of academic language.

Many institutions include in their mission statements priorities like "the first university in . . . ," "number one in distance education," "the largest private university," "the first private university," "the first foundation dedicated to private higher education" and so on. These are also meant to enhance the claim of reliability, while conforming to the relationship that connects academia with tradition.

Most of the phrasings of the mission statements have an underlying tone that lets us detect *a professional-vocational mission*. More than half of the mission statements state this very directly and place it on a prominent position within the text. To quote some typical examples: "our mission is to contribute, along with the public sector, to the creation of specialists competitive on a national and international level," "preparing specialists capable, through their education, to face the demands of a market economy, integrated in the circuit of European political, social, juridical and cultural contacts," "preparing professionals for the Hungarian community that are competitive on an international level, in a Christian spirit," "preparation of specialists for Western Romania," and so on. Many of these institutions also underline the ability of their graduates to fit into national, international, European markets or even "any" employment context, or having the skills that are on demand on the current labor market. The prominent exceptions are to be found in some of the confessional schools and the public-isomorphous institutions that declare their mission as elite education. Nevertheless, they do not explain what they consider to be elite.

While this orientation of studies toward a profession might be generally true for demand compensating higher education anywhere (Levy 1992), on the one hand, it also continues the overall value system imposed on higher education by the former communist regimes across Central and Eastern Europe. I have previously labeled this type of higher education as output-oriented (Reisz 1994). Output-oriented higher education institutions are institutions that identify themselves with and concentrate on their output rather than on the processes that take place within the institution. During the reforms of the communist governments in Central and Eastern Europe, the mission of higher education was set on a systemic level, individual institutions being part of the overall socioeconomic structure. In this context, the mission of higher education was purely vocational. After the so-called polytechnization reforms of the 1950s, the higher education sector was intended to produce the necessary intellectual proletariat for the national economy. All degrees offered by the Romanian Communist higher education system were in fact synonymous with professions. While over half of the graduates were in engineering, all faculties of humanities, arts, and sciences became in-fact teacher-training institutions. This change of focus has been internalized quite well by the population in the half-century of socialism. The academic drift that characterized the development of the public sector after 1990 came as a surprise to a large part of the population. The apparent changes in value system also detected in opinion polls (Reisz 2003a) as well as the expansion of the system and the departure from the professional-vocational character of studies were characteristic of the Romanian higher education sector in the 1990s. Apparent disorder accompanied the autonomization of public higher education institutions. The fear that "society will not be able to handle this large number of graduates" was voiced by the mass media, politicians, and civil society. Therefore, in this context, private universities in need for legitimacy and recognition seemed to prefer to be on the safe side and to present a value system closer to the general opinion of what higher education was meant to do.

On the other hand, as early as 1992, in a series of in-depth interviews on private higher education, Romanian higher education administrators and decision-makers stated that private higher education institutions were created and maintained by the old communist elite (Reisz 1992). There was some empirical truth to it. As a matter of fact, in public higher education, the expansion of the early 1990s lead also to a steep increase in the numbers of teaching staff, a large number of junior faculty entering the system after long years of stagnation. This was also accompanied by a change in the leadership of the higher education institutions. Former leaders, old, prestigious professors were then in the best positions to convert their prestige into money by capitalizing on it on the private higher education market. The need to attract students to private institutions in the early 1990s also resulted in a larger percentage of full professors, well-known names of the higher education sector, and a higher age average of the teaching staff at the private institutions than at the public institutions. Prestigious professors taught the same courses at a number of private institutions. These developments have been reversed by the accreditation law that imposed rules on the teaching staff of all higher education institutions (1993). Nevertheless, as already seen, most private institutions have been created in the years between 1990 and 1992 by academics from the public sector, and their mission statements usually date back to the years of institution building. As a consequence, the vocational mission of these higher education institutions also reflects the internalized values of their creators.

The institutions that try to justify their participation in *elite education* represent a counterpart to the professional-vocational orientation of the schools mentioned above. While the two elements (vocationalism and elite education) are in no respect contradictory, most institutions that mention elite standards or excellence in their mission statements also do not present vocational goals in a prominent position. The elite discourse appears at 6 institutions, all accredited, all based in Bucharest, except one confessional school, all in the public-isomorphous group and having large numbers of students.

These universities have, in fact, the most widespread disciplinary range. One explicitly states being "a university in the classical sense." Here are some more quotations: "an elite university with elite graduates," "an elite higher education institution," "elite university," "devoted to academic excellence."

The quality of the design and the complexity of the Web pages of these institutions make a difference. While there exist no relevant differences between accredited and authorized institutions, elite institutions have relevantly more complex and better designed Web pages.

It should be mentioned that these elite institutions do not intend to present themselves as superior to public higher education, often making the point of being of similar standards or "the same quality as some of the best public universities."

The appearance of elite education in this context could also, to some extent, be explained with the communist heritage. While most of Europe was already approaching mass higher education at the end of the 1980s, in 1990, Romania, as already mentioned, was second but last to Albania with its low number of students per 10,000 inhabitants in all of Europe. Higher education was still elite. The expansion of the overall higher education system, the liberalization of access, and the appearance of private higher education itself threatened this concept. Thus, on the one hand—even if paradoxical—some private higher education institutions, even if intrinsic part of the massification process, had to become, due to the conservativeness of their founders, places for elite education. On the other hand, it can also be considered that the founders of these institutions sensed the need of the prospective students and parents for the elite education that was slowly but surely disappearing in the public sector.

Another important and recurring element, usually mixed with the previous ones, is *the need for new models* for the Romanian youth and society. Confessional, "international," and diversity institutions compete in offering role models for the new generation. One of the major rhetorical patterns of the early 1990s in Romania emphasized the dissolution of the public value system. As communist values were declared obsolete, many politicians, journalists, and other public figures claimed the need for spiritual renewal and the emergence of new role models for youth and society in general. While the communist value system proved to be much more resistant to attacks and its dissolution, more declarative than real, a certain part of it, nevertheless, lost ground. Even if the need for new values to fill the alleged vacuum in the value system was more on the declarative side, suppliers of such offers did appear.

First and most successful of these have been the confessions. As mentioned before, in the years after 1990, eight confessional higher education institutions have been created. In their mission statements, all these institutions refer to their religious affiliations in more or less direct terms. If the Franciscan university simply declares *Deus meus et omnia*,[3] some are more explicit. To take some examples, a Baptist college declares its mission to be to respond to the "calling of the Lord" while another higher education institution "prepares graduates according to the ideals of Christian faith, culture and morals." A Roman-Catholic university claims to "form true shepherds of the souls after the model of our Lord Jesus Christ, the teacher, priest and shepherd."

Another supply of values and models comes from abroad. Romania has a Romanian-American university, a Romanian-British university, and a Romanian-German university. All these institutions are intent to import role models for the younger generation. Probably the most explicit is the mission statement of the Romanian-American university: "our mission is to promote the educational values of American higher education and the behavioral model of American society." A further 8 of the 38 mission statements collected include terms like "international" and "European" to describe their goals, intended standards, or relevance. There exist also a couple of universities that include in their mission references reconsideration of national traditions, one of these even using a nationalist discourse.

The third group of institutions that competes in reforming the values of Romanian society is the diversity group. Of the five institutions included here, three are especially clear in the business of moral redemption: two institutions are dedicated to ecological value systems, while one follows the ideology of Waldorf education. Let us quote from the mission statement of the Waldorf teacher training college: "become a different kind of a teacher—nonconformist, responsible, creative."

Regarding the explicit references to foreign models, there exist implicit references to American models of higher education in some of the mission statements. These are interesting mainly because they are not clarified. It is not that the model is called forth as a legitimation bolster. In these cases, it might be really more of a model in a form of mimetic isomorphism to use the term of DiMaggio and Powell (1991). The references to be mentioned in this context are those to community colleges (or community higher education institutions) that can be found at two of the regional institutions or those offering liberal education (also two cases). In one of the mission statements that mention liberal education, the term "liberal education" is also defined in the sense of the liberal arts colleges in order to avoid confusion with political liberalism, but without mentioning the American model. These situations show that at least some of the institutions are in search of models of private institutions also beyond the discursive level.

Another interesting element is the mentioning of a ban on political activities within the university or of an apolitical status of the institution. As Levy mentions in his classic work on private higher education and the state in Latin America, political involvement in the public sector was one of the reasons for the appearance or the expansion of the private sector (Levy 1986). The ban on political activism within some private universities was also mentioned in the series of interviews with administrators of private higher education institutions conducted in 1992. In the early 1990s, students' movements and strikes plagued public universities. Nevertheless, these were not important or dangerous enough to prompt a movement of the upper and middle class toward the creation of private apolitical institutions. In fact, at the beginning of the 1990s, practically no upper class existed in Romania.

Comparatively few mission statements refer to the internal activities, the processes that characterize the institutions. Only two institutions define themselves in relation to internal factors. The descriptions as "a study community" or "a house of knowledge" are somewhat vague. More common references to processes within the institution relate to the quality of the teaching staff and/or curriculum. Still, these remarks are not prominent in the mission statements, appearing either in comparatively longer statements or in other presentations of the institution.

Conclusions

Thus we have seen that almost all private higher education institutions define themselves in relation to external factors. Private higher education institutions in Romania prove to be externalist as defined by Maurice Kogan in his recent paper (2005). In this definition, there are "socially relevant assumptions resting in social contexts" that justify their form of knowledge, rather than the "internalist perspective relying on prestige of epistemic communities." This finding is also relevant to my previous remarks on the output orientation of private higher education institutions. Private higher education institutions place the legitimation, or justification for the knowledge they claim to create, disseminate, and distribute outside the university itself. The knowledge they pretend to emphasize is not legitimated in the circles of scientific peers but in the social environment of the institution. As a consequence, their rationale is discursively set externally as well.

It is interesting to mention that, by doing so, private higher education institutions fashion their mission statements to a value configuration typical to the communist higher education rhetoric of elite professional education, a configuration that rests on two ideological pillars.

On the one hand, there is the professional-vocational character of higher education that can be traced back to a Napoleonic model of higher education, a model that had its followers in Romania as early as 1818. This materializes in the role of higher education to "prepare specialists," its inclusion in economic and administrative rather than cultural or scientific flows, its relation to "needs of the economy," "real needs of the labour market." What has been rarely said, but implied in this rhetoric, is the fallacy that market needs are stable, or else can be planned and evaluated.

On the other hand, communist higher education subscribed to an elite model of higher education. This, in the Romanian case, is clearly visible in the very low participation rates of higher education up to 1990. The basis for this elite understanding of the higher education system lies in the belief that there exist little natural, genetic talents that limit the possibility for higher study and that they are so scarce in the population that an increase in higher education participation is not possible or beneficiary. The undertone of this remark is that the benefits of higher education are viewed on a systemic level and the benefits of higher education for the individual seem to be considered as irrelevant, or irrelevant on systemic level. These types of discourses are in no way limited to Romania, and other former communist states, the example of recent German debates on higher education being at hand.

Private higher education institutions prove to be mostly conservative in their mission statements as well as in the design of their Web pages. This conservativism relates them not to the traditional meaning of the university, but to the vocational-polytechnic higher education institution of the communist rhetoric of usefulness. Whether or not it is because of their need to satisfy expectations of their stakeholders or normative characteristics of their leaders, these institutions mostly go the safe way of conforming to the value configurations that emerged in the last half of a century. Nevertheless, it should also be stated that the disciplinary structure of private higher education is consistent with and promoter of such professional-vocational higher education missions.

Witnessing the present quest for legitimacy of Romanian private higher education institutions, we find a diverse landscape of institutional discourse. The overall image is of organizations struggling to gain recognition as useful and academic, and conforming to the predominant social understanding of higher education in Romania, where useful translates into vocational and academic into elite.

Notes

1. Institutions are authorized according to a series of quantitative indicators included in a self-evaluation and an evaluation of a visiting expert group. Accreditation can be achieved if a majority of each of the first three generations of graduates pass license examinations at accredited institutions. The proposals of the NCAEA are transferred to Parliament; the final accreditation takes place by law.
2. Quotation marks are used to distinguish these institutions from institutions that are indeed international, as, for instance, the Central European University in Budapest. In this case, it is the legitimation discourse that is "international," rather than the institution.
3. In English: "My God and My All," the motto of St. Francis.

Bibliography

Bowles, S. and Gintis, H. 1976. *Schooling in Capitalist America*. New York: Basic.

Berger, P. and Luckmann, T. 1966. *The Social Construction of Reality*. New York: Doubleday.

Dima, A. 1998. Romanian private higher education viewed from a neo-institutionalist perspective. *Higher Education in Europe*, Vol.3. Bucharest: CEPES UNESCO.

DiMaggio, P. and Powell, W. 1991. The Iron Cage Revisited: Institutional Isomorphism and Collective Rationality in Organizational Fields. In P. J. Di Maggio and W. W. Powell (eds.), *The New Institutionalism in Organizational Analysis*. Chicago and London: University of Chicago Press, pp. 63–82.

Geiger, R. 1986. Finance and function: Voluntary support and diversity in American private higher education. In D. C. Levy (ed.), *Private education. Studies in choice and public policy. Yale studies in non-profit organisations*. Oxford: Oxford University Press, pp. 214–237.

Hannan, M. and Freeman, J. 1989. *Organizational Ecology. Cambridge*. Massachusetts: Harvard University Press.

Ladányi, A. 1991. *A Felsöoktatás Összehasonlitó Sztatisztikai Elemzése Comparative Statistics in Higher Education*. Budapest: HIER.

Levy, D. 1986. Higher *Education and the State in Latin America. Private Challenges to Public Domination*. Chicago: University of Chicago Press.

——— 1992. Private Institutions of Higher Education. In Burton Clark and Guy Neave (eds.), *The Encyclopedia of Higher Education*.Pergamon Press, Oxford, pp. 1183–1194.

——— 1999. When Private Higher Education Does Not Bring Organizational Diversity: Argentina, China and Hungary. In P. G. Altbach (ed.), *Private Prometheus: Private higher education and development in the 21st century*. Westport, Conn.: Greenwood Publishing Co, pp. 17–50.

——— 2005. The Unanticipated Explosion: Private Higher Education's Global Surge. *Comparative Education Review*, 50, pp. 217–240.

Kogan, M. 2005. Modes of Knowledge and Patterns of Power. *Higher Education*, Vol. 49.

Manning, P. 1992. *Organizational Communication*. New York: Aldine de Gruyter.

Meyer, J. 1977. The Effects of Education as an Institution. *The American Journal of Sociology*, Vol. 83, No. 1, pp. 55–77.

Parsons, T. 1956. Suggestions for a sociological approach to the theory of organisations. *Administrative Science Quarterly*, Vol. 1, pp. 63–85.

Pellert, A. 1999. *Die Universität als Organisation. Die Kunst Experten zu managen*. Wien: Böhlau.

Reisz, R.D. 1992. Private Higher Education. Romania. *Education*, Vol. 2. Budapest: HIER.

——— 1994. Curricular patterns before and after the Romanian Revolution. *European Journal of Education*, Vol. 3, pp. 281–290.

——— 1997. Private higher education in Romania. A second look. *Tertiary Education And Management*, Vol. 3, No.1, March, pp. 36–43.

——— 2003a. Hochschulpolitik und Hochschulentwicklung in Rumänien zwischen 1990 und 2000 (Higher education policy and higher education development in Romania between 1990 and 2000), HoF Arbeitsberichte 1/2003,

Wittenberg: Institut für Hochschulforschung, Martin Luther Universität Halle Wittenberg (Institute for Higher Education Research at the Martin Luther University Halle Wittenberg).

——— 2003b. Public policy for private higher education in Central and Eastern Europe. Conceptual clarifications, statistical evidence, open questions, HoF Arbeitsberichte 2/2003, Wittenberg: Institut für Hochschulforschung, Martin Luther Universität Halle Wittenberg (Institute for Higher Education Research at the Martin Luther University Halle Wittenberg).

Sadlak, J. 1994. The Emergence of a Diversified System: The State/Private Predicament In Transforming Higher Education In Romania. *European Journal of Education*, Vol. 1, pp. 13–23.

Scott, P. 2000. Higher Education in Central and Eastern Europe: An Analytical Report. In ten years after and looking ahead: a review of the transformations of higher education in Central and Eastern Europe. *Studies in Higher Education.* Bucarest: CEPES.

Slantcheva, S. 2004. Legitimating the difference: Private higher education institutions in Central and Eastern Europe, paper presented at the Sofia International Workshop *In Search of Legitimacy: Issues of Quality and Recognition in Central and Eastern European Private Higher Education.* June 20–21, Sofia: Bulgaria.

Suchman, M. 1995. Managing Legitimacy: Strategic and institutional approaches. *American Management Review*, Vol. 20, No. 3, pp. 571–610.

Suddaby, R. 2002. Rhetorical strategies of legitimacy: Vocabularies of motive and new organizational forms. Paper presented at the 18th EGOS Colloquium, Barcelona, Spain, July 5–7.

Whitehead, J. 1977. The origins of the public-private distinction in American higher education: A story of accidental development. *Higher Education Research Group.* New Haven, CT: Yale University Press.

Chapter Seven

Between the State and the Market: Sources of Sponsorship and Legitimacy in Russian Nonstate Higher Education

Dmitry Suspitsin

Introduction

A major development in Russia's post-Soviet higher education has been the emergence and proliferation of private higher education institutions. Typically referred to as nonstate for a number of important reasons,[1] these institutions have profoundly altered the organizational landscape of Russian higher education and have considerably expanded the capacity of the system to provide services to various segments of the public (Solonitsin 1998). After the 1992 legislation (Federal Law 1999) introduced the term "private educational institutions" and permitted the operation of nongovernmental forms of higher education, this sector has experienced rapid and robust growth, having mushroomed to over 400 institutions. It currently makes up roughly 38 percent of all 1,071 higher education institutions, serving about 15 percent of all students in the country (Center for the Monitoring and Statistics of Education [CMSE] n.d.).

Despite this vibrant growth, the sector's societal recognition, or legitimacy, is marked with controversy stemming from different assessments among its relevant constituencies and stakeholders. Emerged in conditions of a nascent market economy and cultural pluralism, nonstate institutions orient themselves toward meeting the exigencies of new market arrangements and social values. They tend to closely coordinate study programs and student learning outcomes with the customers' and employers' pragmatic demands, finding considerable acceptance and support through expediency and practical value in the marketplace. In contrast, the government mainly assesses institutional appropriateness and bestows its recognition through a formal process of state accreditation, emphasizing traditional standards of academic quality and professionalism. Given these different (and often divergent) expectations for social acceptance, many nonstate institutions find themselves between a rock and a hard place when trying to adapt to the legitimacy pressures from the state and the market. How Russian private higher education institutions deal with the legitimacy challenges posed by the state and the market is the primary focus of this chapter.

As in most post-communist societies, the proliferation of the Russian nongovernmental forms of higher education and the problem of their legitimacy are linked to broader socioeconomic processes of the transition from centralized state control to a free market system, which typically occur in conditions of conflicting and contested regulatory and normative frameworks (Levy 2004; Lewis et al. 2003). The introduction of full-scale free market reforms in Russia in the 1990s diminished the authority of the state with respect to its financing and governance controls and diversified the sources of sponsorship and legitimacy available to higher education, particularly for newly emerged nonstate institutions. Indeed, whereas the Soviet system allotted the state an almost exclusive role of sponsor, provider, and regulator of higher education, the new policies of decentralization, deregulation, and privatization[2] allowed for the emergence of multiple and diverse entities of sponsorship and provision of higher education services: the state, individual consumers, and the nascent sectors of business and civil society. As a result, nonstate institutions found themselves adhering to ambiguous standards and rules and orienting toward multiple and often competing legitimation sources.

This study draws a general picture of constituencies and stakeholders that Russian private higher education employs as sources of legitimacy. Using institutional founders as indicators of the sources of sponsorship and legitimacy available to nonstate institutions at the time of their establishment, this chapter lays out several distinct legitimacy-building orientations of nonstate universities[3] and offers partial explanations for the success of these strategies.

I group and examine nonstate institutions according to their founding entities. The choice of founders as a criterion of institutional differentiation serves several purposes. Institutional founders are extremely important to new organizations in their legitimacy building. Not only do institutional organizers often determine the identity of new ventures (Stinchcombe 1965), but they also directly show which organizations and actors at founding support and endorse the de novo enterprises. The assumption is that the composition of founders may point to the initial sources of legitimacy and legitimation orientations of newly established nonstate institutions.

Ranging from governmental organizations to state universities to purely private actors, the founders with their various configurations in effect represent a continuum of privateness[4] and point to varying degrees of proximity to either state-run or private organizations on the part of nonstate universities.

Overview of Private Higher Education in Russia

Research on worldwide private higher education has documented three major reasons for the emergence of private initiatives in education: (1) religious or other cultural purposes of various groups wishing to promulgate their values; (2) provision of elite alternatives to public higher education; and (3) compensation for the inability or unwillingness of the state and the public sector to meet the demand for higher education (Geiger 1986, 1991; Levy 1986; Lewis et al. 2003; Reisz 2003). For example, Levy (1986) notes these three rationales in his seminal study of private higher education in Latin America categorizing institutions into value-centered, elite, and demand-absorbing. In another influential study, Geiger (1986) similarly observes that the heterogeneity of demand for private higher education may take three forms, including *more* higher education to meet the excess general demand; culturally *different* kind of higher education to address cultural and religious pluralism; and *better* higher education to meet the demand for quality (mainly research oriented) education. In Russia, the failure of the Soviet state gave rise to heterogeneous and immediate demand for more, different, and better higher education, leading to the emergence of Levy's three institutional types at the same time. The majority of nonstate institutions, however, were set up to respond to a massive demand for new, market-oriented education (Solonitsin 1998).

As indicated earlier, the conditions for the emergence of nonstate universities in Russia were created by the initiation of free market reforms in late 1980s and early 1990s (Kirinyuk et al. 1999). The newly emerged economic institutions of a market economy required an education that the state sector of higher education was not able to meet effectively. With its inertia and major deficiencies relative to the needs of a market economy, including neglect for such disciplines as economics, management, law, and sociology, as well as little emphasis on adult education and retraining, the system of state universities initially produced human capital of little value for a new economic order. In these circumstances, the emergence of private universities with their lead in addressing the market needs was a force that helped enhance societal adaptation in the changing socioeconomic environment (Etzioni 1987).

The proliferation of Russia's private sector generally follows the patterns that occurred in other parts of the world, with its surprise element of growth, lack of planning on the part of policymakers, initial explosion and subsequent stagnation in growth, and other common characteristics globally (Levy 1992, 2002). The precursors of private higher education institutions appeared and gained momentum in the absence of any solid legal framework in the late 1980s, in effect functioning as for-profit enterprises that offered various professional development and certificate programs (Veniaminov 2002). The unexpected vibrant growth of private initiatives in education, along with an organizational crisis within the state sector of education, compelled the government to impose regulation and accountability upon the rapid and somewhat chaotic developments in education. After the 1992 legislation (Federal Law 1999) provided for the founding and functioning of educational institutions of various legal organizational forms, including state, municipal, and nonstate, and indeed introduced the term of private educational institutions (albeit without the term's clear-cut definition), the private sector became the fastest growing segment of the higher education market with respect to the number of institutions. Of the 451 institutions that opened their doors in the 1990s, 80 percent were established in the private sector (Center for Research and Science Statistics (CRSS) 2002). Table 7.1 details the expansion of Russian higher education between 1993 and 2004.

Table 7.1 Higher education institutions and student enrollment in Russian higher education, 1993–2004

Academic year	*State institution*	*Enrollment in state institutions*	*NonState institution*	*Enrollment in nonstate institutions*	*Total higher education institutions*	*Total enrollment in higher education*
1993/94	548	2,543,000	78	70,000	626	2,613,000
1994/95	553	2,534,000	157	111,000	710	2,645,000
1995/96	569	2,655,000	193	136,000	762	2,791,000
1996/97	573	2,802,000	244	163,000	817	2,965,000
1997/98	578	3,047,000	302	202,000	880	3,248,000
1998/99	580	3,347,000	334	251,000	914	3,598,000
1999/00	590	3,728,000	349	345,000	939	4,073,000
2000/01	607	4,271,000	358	471,000	965	4,741,000
2001/02	621	4,797,000	387	630,000	1,008	5,427,000
2002/03	655	5,229,000	384	719,000	1,039	5,948,000
2003/04	654	5,596,000	392	860,000	1,046	6,456,000
2004/05	662	5,860,000	409	1,024,000	1,071	6,884,000

Source: Center for the Monitoring and Statistics of Education (CMSE), Ministry of Education and Science, n.d. Retrieved from http://stat.edu.ru/stat/vis.shtml

After over a decade of impressive growth, the private sector has emerged as a vibrant and diverse component of the higher education system effectively competing with state universities in market-related fields of study and often offering programs in areas unavailable in the state sector. In addition to the general entrepreneurial, market-oriented role that a sizeable proportion of Russian private institutions perform, a few institutions (3 or 5), with support from international foundations and foreign universities, have assumed an elite academic role of offering genuinely Western-style academic programs (particularly in economics), and many others have pursued distinctive, value-centered roles. In the last category mentioned, private institutions cater to the populations previously neglected under the Soviet regime, such as various religious groups and ethnic and other cultural minorities, offering programs in theology and cultural theory, special education and psychology, humanities and social sciences that were either nonexistent or underemphasized under the Soviet model of education.

As of 2005, 409 private institutions[5] (as compared with 662 state-run ones) account for roughly 15 percent of enrollments in higher education (CMSE n.d.). Generally located in metropolitan and large urban centers, these institutions offer mainly market-related programs in economics, law, psychology, sociology, social work, business administration, and other disciplines that do not require much investment in equipment and research infrastructure. Their generalized profile includes responsiveness to the needs of the labor market, flexibility of the course offerings and curricula, frequent use of learner-centered instructional methods, heavy reliance on part-time faculty, tuition dependence, loose admissions requirements, limited concern about research, and other features typically ascribed to private institutions worldwide (Altbach 1999; Geiger 1986; Levy 1986, 1992). Without direct subsidies from the central government, and despite their nonstate status, a great number of Russian private institutions nevertheless remain closely connected to various governmental bodies (particularly local administrations) and state-owned resources[6] (Volkov et al. 2004). In fact, many nonstate institutions are founded by state universities and by governmental organizations taking advantage of their founders' state-owned assets or administrative connections.

Current law provides for the founding of nonstate institutions by organizations, private individuals, or their combination. Typical founders and owners include government authorities; state universities, research institutes, and centers; national and foreign businesses and industrial enterprises; international and private foundations; religious organizations; and Russian and foreign citizens (International Finance Corporation 1998). As nonprofit organizations by statute, nonstate institutions are only allowed to generate revenues that should support solely the operation of educational activities.

A Theoretical and Conceptual Framework

Organizational legitimacy is a central concept in organization theory. It is viewed as a prerequisite for the survival of organizations. According to Suchman, legitimacy is "a generalized perception or assumption that the actions of an entity are desirable, proper, or appropriate within some socially constructed system of norms, values, beliefs, and definitions" (1995, p. 574). Organizational legitimacy generally presupposes an organization's acceptance as appropriate, trustworthy, and worthwhile by its constituencies and stakeholders who approve of, endorse, or support it.

Scholars of higher education have identified social systems' major centers of power, which lay demands on, regulate, and endorse higher education institutions. For instance, Clark (1983) distinguishes among three major forces that coordinate higher education systems: state authority, the market, and academic oligarchy. State coordination refers to a framework of rules designed to "steer the decisions and actions of specific societal actors according to the objectives the government has set and by using instruments government has at its disposal" (Neave and van Vught 1994, p. 4). Market regulation that arises from mutually beneficial self-interested relationships is conceived as a method of structuring behavior and managing interdependencies among the actors in an exchange. Academic oligarchy refers to the coordinating agency of "academic guilds" to guide decisions and behavior in higher education (Clark 1983). Positioned within a triangle, with each corner representing the extreme manifestation of one force and marginal influence of the other two, higher education systems may be subject to these pressures in varying degrees in different national contexts.

Clark's triangle of coordination in effect represents major sources of legitimacy and sponsorship available to higher education institutions in the external environment. Coordination or regulation on the one hand and legitimation on the other hand are interconnected processes to the extent that adjustment to regulatory controls is an organizational mechanism effecting an organization's legitimation (Scott 1998): following frameworks and rules signals to stakeholders adherence to expectations of appropriate behavior that is rewarded with perceptions of legitimacy and with other assets.

To further specify the constituencies and stakeholders in higher education who confer legitimacy within the domains of the state, the market, and the academic sphere, it is important to supplement Clark's model with DiMaggio and Powell's notion of "organizational field" defined as a community of organizations with some functional interest in common, including "key suppliers, resource and product consumers, regulatory agencies, and other organizations that produce similar services or products" (1983, p. 148). While retaining Clark's entities of state authority and the market intact, I replace the academic oligarchy element with a notion of "higher education community" to include and account for a critical legitimizing role of peer organizations (particularly state universities) in the Russian context.[7] I envision Clark's three sources of influence as three overlapping circles, rather than as a triangle, to emphasize the idea of hybrid organizational arrangements and the interaction among the legitimating entities in a higher education field.

This tripartite framework can be effectively applied to the higher education context in Russia (and in several other post-communist countries) where the role of the state (albeit diminished) remains influential, market forces are increasingly exerting more regulation, and the influence of public higher education over its private counterpart is pronounced. Each major realm of sponsorship and legitimacy consists of a set of constituencies and stakeholders who are capable of conferring legitimacy. In Russian higher education, the state authority is represented by the Ministry of Education and Science and its accreditation agency, other ministries who run their higher education institutions, as well as various central, regional, and municipal administrations that have some control or sponsorship of higher education centrally or locally. The market dimension in Russia includes student clientele, employers, sponsoring business and industry organizations, and privately owned, entrepreneurial educational organizations. The academic organizational community comprises institutions whose legitimacy is well established, namely state-supported universities and specialized research institutes and academies, including the Russian Academy of the Sciences and the Russian Academy of Education.[8]

Major actors in the Russian higher education field confer legitimacy of different essence[9] through two principal mechanisms. One process is central coordination exerted by the state with its formal powers. The state grants legitimacy by top-down supply-side regulation of academic quality through the imposition of national standards for higher education at accreditation.[10] Effected by "coercive isomorphism" (DiMaggio and Powell 1983), legitimacy in this case mainly means conformity with laws and regulations and compliance with what Clark (1983) calls "competence" values, or adherence to academic excellence.[11] Accredited institutions bear the state's hefty seal of authorization of institutional trustworthiness and merit that facilitates their graduates' acceptance in society.

The other mechanism is coordination by mutual adjustment of interests and reciprocal offers of benefits (Geiger 2004; Lindblom 2001), effecting legitimacy termed "pragmatic" by Suchman (1995). Consistent with Clark's (1983) values of "liberty" that embrace choice, initiative, diversity, and innovation in higher education, this form of legitimacy rests on the self-interested calculations of universities' stakeholders and constituencies who approve of and lend their support in exchange for some practical benefits that universities bring them. Endorsement and support expressed through exchange-based relationships largely occur in the marketplace (albeit not necessarily) that generally legitimates through its fundamental regulatory processes of supply and demand and resource exchanges. Higher education institutions are legitimated in the process of reciprocal conferment of benefits when their enrollments grow or when employers hire their graduates. In the Russian context, exchange-based legitimacy also comes from various partnerships between government organizations (e.g., local administrations) and universities or between higher education institutions themselves (e.g., partnerships between state and nonstate universities). The latter arrangements often arise from inherent and interconnected market forces of competition and cooperation in the higher education field at large, including markets for students and faculty (Geiger 2004).

In general, legitimation is largely evident when influential social actors and exchange partners contribute resources to universities thereby expressing their endorsement and signifying and acknowledging the legitimacy of the receiving institutions[12] (Hybels 1995). As DiMaggio explains, "subsidiary actors provide legitimacy to the new organizational form by providing resources that render its public account of itself plausible" (1988, p. 15).

Method

Data and Procedures

This study draws on institutional data from the Russian Ministry of Education and Science (CMES n.d.). Out of a total of 369 private institutions in the database, 308 cases were included in the analysis. With founding entities as the primary variable of institutional differentiation, analysis was conducted yielding several patterns of nonstate university founding. Institutions were coded and assigned to one of the four groups established by (1) private businesses and educational entrepreneurial organizations, (2) private individuals, (3) state universities and research institutes as sole founders, and (4) multiple and hybrid entities. Cross-tabulated descriptive statistics were used to compare the groups on indicators of legitimacy. In classifying nonstate institutions' orientations toward the realms of the state, market, or higher education community, several essential attributes of institutional founders were considered. The criteria of differentiation and the coding scheme[13] were based on whether the founders were organizations or private individuals; educational or noneducational organizations; state-related or independent, nongovernmental structures; for-profit or nonprofit organizations; and diverse or homogeneous entities.

Legitimacy Indicators

Accreditation and Ministry Rankings

Shares of accredited institutions and Ministry rankings may be viewed as indicators of state-granted legitimacy signifying an endorsement by the state. Accreditation is arguably the most comprehensive and important indicator of state-imposed legitimacy, meaning an institution's right to award degrees whose quality is recognized and therefore endorsed by the state. To private higher education institutions, accreditation is particularly important because, in addition to quality implications, it facilitates societal acceptance in the public sphere through conferring a number of social privileges, including student deferment of military service, ability to occupy positions in governmental organizations after graduation, and a right to enter graduate school at state universities (Solonitsin 1998). Under review at accreditation are many aspects of teaching, research, and service, including student achievement; qualifications, research productivity, and professional involvement of the faculty; funding and facilities for research; student services and social facilities; instructional resources, including libraries and information technology facilities; financial viability (per student expenditures); internal quality assurance of academic programs (internal assessment and evaluation); and job placement rates of graduates.

Similar to accreditation, Ministry rankings[14] are another certification mechanism (Rao 1994) signifying an endorsement from the state. The evaluation criteria for the rankings and accreditation overlap considerably, taking into account mostly institutional input characteristics and some performance indicators (e.g., faculty research projects and publications).

Graduate Programs and Physical Plant Infrastructure

Graduate programs and a physical plant infrastructure may be taken as indications of the legitimizing influence of the academic realm and the market. To the extent that graduate education implies engagement in research, institutions offering graduate programs may be accorded a high status. The "size of physical plant"[15] either owned or rented by institutions may point to their relative success in resource acquisition. As mentioned earlier, legitimacy may be indirectly assessed and is often evident through the flow of resources,[16] that are not only necessary for economic fitness but also provide evidence that key constituencies and stakeholders view organizations' activities legitimate. Owning or renting adequate instructional and supporting facilities on the part of Russian private higher education institutions may imply the support garnered from influential social actors through beneficial exchanges.

Limitations and Ambiguities

The study's reliance upon the data from the Ministry entails limitations. The database includes only formal indicators of state-regulated legitimacy and lacks data for ascertaining the legitimizing role of market coordination (e.g., job placement rates). As a result, major legitimacy indicators in the study may not only make some groups of institutions figure more prominently than others, but they also may shift maximum emphasis toward the legitimizing role of the state and state-granted legitimacy.

Ambiguity arises when using accreditation as a legitimacy indicator. Because of its comprehensive framework of evaluation, accreditation also may partially gauge legitimation effects of the academic community and the market in addition to its core instantiation of the state authority and state-granted legitimacy. For example, the standards for higher education that accreditation is intended to ascertain have been developed by the professional community and academe. Similarly, accreditation partially gauges institutions' market prowess through an indication of employers' formal requests for graduates submitted to institutions, as well as through measures of resource acquisition that largely occurs in the marketplace (e.g., markets for students and faculty, etc.).

Findings: Institutional Group Comparison

Analysis of the founding entities yielded four distinct groups of nonstate higher education institutions, revealing a multifaceted and complex structure of institutional founding. The founders represent a diverse group of entities running the gamut of private and public, for-profit and nonprofit, academic and nonacademic, and state-run and independent organizations. They include private individuals, nonprofit or for-profit companies, organizations, associations, and foundations, as well as governmental organizations and agencies, and state and nonstate academic institutions. The four groups of institutions are located in different places within Clark's triangle and have different degrees of privateness and of proximity to the state sector.

Group I (*Private Proper*) consists of institutions founded by independent entrepreneurial organizations, including businesses and privatized industrial enterprises, early private educational enterprises started before the 1992 law on education, and viable private higher education institutions that established other private institutions often as franchisers after several years of successful operation. Among other groups, this set of institutions is the most *private* in its founding structure and most independent from the state with respect to resource management. It represents orientation toward the marketplace as the dominant legitimation strategy.

Group II (*Person-Only Founding*) includes institutions founded by one or several private individuals without formal involvement of organizations. The backgrounds of these individuals are diverse, varying in the extent of closeness to either private or state-run organizations and encompassing entrepreneurs from the business community, top-level administrators and researchers from state universities and specialized research institutes, foreign and domestic individuals affiliated with religious or civil organizations, and others. Although this group includes many private entrepreneurs with business backgrounds and may appear to represent purely private initiatives, its founding structure is somewhat ambiguous due to the difficulty of assessing the kinds of supporting organizations behind the founding individuals. For example, many individuals are former or present rectors of state universities, maintaining close ties to the state sector of higher education. Also, institutions founded by religious leaders imply the backing of their value-based organizations rather than those oriented toward the marketplace.

Group III (*State University Proximate*) comprises institutions established by one or two state universities exemplifying the prominence of state higher educations as critical resource holders and legitimating entities. Proximity to state universities allows this group to take advantage of their resources and to assume a veneer of state university crucial to legitimacy perceptions.

Group IV (*Hybrid, Multiple-Source*) institutions include multiple and diverse founding entities, embracing different configurations of actors from the previous three groups and additionally from various government organizations, such as local, regional, and central administrations and ministries. A key feature of this group is that co-founders are always multiple and heterogeneous composed to a different extent of audiences from two or all three of Clark's centers of sponsorship. For example, an institution from a multiple-source group may have the following combination of cofounders: a state university, a city administration, a private university, and a business. This group represents hybrid founding arrangements and interpenetration of the realms of the state, market, and higher education community.

Institutional group comparison on legitimacy indicators reveals a higher stature of institutions established by multiple and diverse entities and by state universities relative to institutions representing independent private initiatives. The multiple-source and state university proximate groups have significantly higher shares of accredited and ranked institutions, as compared with the private proper and person-only groups. Approximately nine out of ten institutions in the hybrid group and three out of four institutions in the state university proximate group are accredited by the state, while only two in three institutions enjoy accreditation in the remaining groups. Similarly, approximately one in five institutions in these groups enjoys the Ministry ranking, as compared with one in ten ranked in the private proper category. Table 7.2 presents the study's descriptive statistics.

Hybrid institutions are leaders in graduate education among the peers, whereas the scope of graduate education in institutions established by private initiatives is only marginal. Indeed, multiple-source institutions account for 68 percent of all graduate students in the private sector. Institutions that are offshoots of state universities also compare favorably on this indicator with institutions founded by private businesses and individuals. However, the comparison would be even more advantageous for state university spin-offs if the data accounted for a major advantage derived from their affiliation with parental universities, namely the ability of their students to enroll in graduate programs of the founding universities.

Table 7.2 Descriptive statistics on Russian nonstate higher education institutions grouped by founders, 2003

	Group I private proper	*Group II person-only*	*Group III proximate to state universities*	*Group IV hybrid, multiple-source*	*Private higher education average*
No of institutions	64	105	70	69	—
% Accredited	66	66	76	88	73
% of Institutions ranked	11	12	21	19	16
% of Institutions with graduate school	5	3	6	10	6
Average no of graduate students per institution	1	1	4	20	6
% of Graduate students among all graduate students at nonstate institutions	4	5	13	68	—
Area of physical plant per student (in sq. m)	3.9	3.7	4.5	5.5	4.2
No of headcount students per institution	1,540	1,740	1,350	1,660	1,600

Source: Center for the Monitoring and Statistics of Education (CMSE), Ministry of Education and Science, n.d. Retrieved from http://www.edu.ru/db/cgi-bin/portal/vuzp/vuz_sch.php

The measure of physical plant infrastructure also points to Group III and Group IV institutions' higher levels of social support. For instance, an average area of the hybrid group's instructional and supporting facilities per student is considerably larger than that of institutions representing purely private initiatives. This same indicator for state university spin-offs is also larger than that for the institutions established by private businesses and individuals. Still a major advantage of state university satellites is the fact that they enjoy access to facilities of their parent universities, which would further bolster their case if reflected in these data.

Discussion and Conclusions

The structuring of Russian private higher education during the 1990s occurred against the backdrop of a turbulent socioeconomic environment with its ambiguous normative and regulatory frameworks, a rapid and unanticipated surge in private initiatives in education and in society at large, and acute and widespread societal debate about the private sector's legitimate roles and goals. The conditions surrounding this sector's development, such as legal indeterminacy and rapidly changing socioeconomic normative order left an indelible imprint on these institutions' mode of identity construction and legitimacy acquisition. Functioning in the presence of *old* and *new* logics of organizational behavior, the private sector had been challenged to adhere to uncertain institutional standards and to orient itself toward multiple legitimating entities in the realms of the state, market, and higher education community.

Russian private higher education institutions manage to effectively seek the support from both the state and the market in the acquisition of sponsorship and legitimacy.[17] On the whole, market orientation is an essential characteristic of almost all nonstate institutions. Offspring of a nascent market economy, most institutions tailor their academic programs to the needs of the customers and employers, finding sponsorship and legitimacy in the marketplace. Without direct funding from the central government, the marketplace is their prime survival arena. Yet with this backbone modus operandi, nonstate institutions also covet the state's legitimizing blessing in the form of accreditation as many relevant constituencies and stakeholders deem it pivotal to Russian higher education (Solonitsin 1998).

Legitimacy from both the state and the market has limitations. A necessary condition for survival in the long run, state accreditation is not a sufficient condition for uncontested recognition in society at large. To enjoy unquestionable acceptance in society, it is also essential to acquire a cognitive state of taken-for-grantedness and often moral propriety from public opinion, peer organizations, and other social collectivities (Suchman, 1995). Additionally, the state's supply-side regulation of academic quality is sometimes at variance with the demand-driven imperatives of the labor market. On the other hand, the market in Russian higher education is a relatively young social phenomenon experiencing obstacles in its institutionalization due to path dependent cultural reasons (e.g., statist tradition of social organization). Coupled with challenges of a developing market, such as imperfect information and uneducated consumers, the market itself as a legitimating institution may not be sufficient to confer solid, uncontested legitimacy (Solonitsin 1998). These limitations determine to some extent institutions' multiple orientations toward resource and legitimacy acquisition.

This study's findings point to three key orientations of Russian nonstate institutions toward legitimacy acquisition, each drawing on its major adaptive advantages created at founding and tapping in different degrees into legitimacy sources in the realms of the state, market, and higher education community. Some institutions build legitimacy by diversifying their base of supporting social actors and by co-opting multiple influential audiences, particularly local governmental organizations, into their operations; others cling to state universities for resources and legitimacy while at the same time exploiting opportunities in the marketplace; and still others find legitimacy largely in the private realm of the market.

The evidence suggests that the first two orientations are more effective when institutions are particularly concerned with state-granted legitimacy. Nonstate higher education institutions with multiple and diverse ties to prominent entities or with ties to legitimate organizations holding critical resources, such as state universities, maintain higher levels of social acceptance from the state than private institutions lacking such ties for a number of reasons: their legitimacy base is more diverse and solid and their close contact with the state sector and access to state-owned resources enable them to capitalize on their relationships more freely. Like a chameleon changing color in different environments, institutions of hybrid origin may assume different roles or appearances when appealing to various types of constituencies and stakeholders. Institutions mimicking state universities largely develop through symbiosis, benefiting from the reputations and resources of their parent universities.

Indeed, institutions with multiple and heterogeneous ties to influential social actors have major advantages. Their multi-source orientation toward diverse and powerful audiences, including state universities, municipal and regional administrations, and the business community, provides them with dense and diversified interorganizational networks thereby enabling them to tap into a large resource and legitimacy base in the higher education field. The participation of governmental structures, especially local and regional administrations, in hybrid educational arrangements is particularly noteworthy. While some of these governmental organizations lend symbolic support that enhances publicity, others bring tangible assets by providing preferential access to physical plant and buildings (often conveniently located in desirable downtown areas) and by offering rent at reduced prices and relief from local taxes. In some cases, connections with government officials also facilitate obtaining licenses for educational activities and accreditation.

Nonstate institutions established by state universities also fare well with respect to recognition from the state. Their relative success is contingent upon mutually beneficial relationships with their parent universities. These relationships, however, vary in the kind of interaction and in the extent of influence of the founding state universities over governance affairs of the satellite private counterparts. Although separate statutory bodies legally, many nonstate offshoots are often housed within state university premises, employ many of the same faculty members, and share many resources of the founding state universities, including libraries, sporting facilities, dormitories, and research laboratories. In these cases, nonstate institutions are frequently governed informally by state university rectors and in effect operate as branches of state institutions. In other cases, institutions of this type are administratively and financially independent, forging legitimacy through partnerships with their parental state universities.

In the early 1990s, this arrangement was a critical factor that enabled both institutions to survive during a time of financial hardship. Motivations to start nonstate institutions on the part of state universities varied from the desire to have closely tied organizations as "cash generation engines" for the founding institutions (Tomusk 2002) to the idea of creating independent higher education institutions to promulgate innovations that were greatly circumscribed by the inertial organizational structures of the state sector of higher education. Additionally, these newly established institutions also enabled state universities to keep many of their faculty members and researchers from leaving positions at state-run institutions by offering additional part-time employment in the private sector in the circumstances of extremely tight budgets (Kirinyuk et al. 1999). In return, private institutions enjoyed the benefits of association with prestigious academic organizations and access to their resources that facilitated their legitimacy acquisition.

State university spin-offs have one central advantage that has helped them to develop and to cope with legitimacy threats. At founding, they were allied with state-run academic institutions that had all the resources necessary for starting a private institution in abundance: organizational knowledge of how to organize teaching and learning processes and practices, qualified faculty members eager to participate in new enterprises, skilled leaders and administrators, and classroom and other facilities.

From the standpoint of transaction cost economics (Williamson 1985), these partnerships are viable because they offer the participating parties an advantage of low transaction costs that stem from low uncertainty and low risk of exchange-based relationships. Indeed, uncertainty of a transacting relationship between the founding university and the satellite private institution is low because of a high level of trust resulting from close interpersonal ties of the people involved.

The study's data indicate that institutions orienting themselves solely toward the marketplace are subject to legitimacy threats from the state to a greater extent than their counterparts employing additional sources in other realms of the Russian higher education field. Yet, organizations generally do not need the support of all the segments of society to remain legitimate. As Pfeffer and Salancik (1978) note, a legitimate organization can be endorsed by a segment of society large enough to ensure its survival in the face of adverse reactions from some social groups. Precisely because of the multiplicity of legitimizing audiences with their diverse values espoused, private higher education institutions may resort to the kind of social support that Levy in this volume calls "niche legitimacy," drawing on the interests and sponsorship of a narrow segment of society that enables them to remain socially and economically fit.

To generalize, legitimacy assessments of Russia's private sector of higher education presently remain ambivalent. After over a decade of operation, private higher education institutions are still questioned with regard to their trustworthiness and reliability by various actors, particularly the state authority. Since mid-1990s, the government has been following a course of tightening regulations and raising accreditation requirements for the private sector. In addition to increasing targets for full-time faculty employment and per capita space of instructional facilities, it has recently devised an *index of economic viability*, with an eye on the private sector. The index establishes a minimum level of annual per student expenditures in various specializations at higher education institutions necessary for obtaining accreditation (Ministry of Higher and Professional Education 2001). For example, for a group of economics or management majors of 25 students, annual costs of education are set in the range of roughly US $11,000 and US $21,670, amounting to US $440 per student annually. To open a new program, an institution needs to demonstrate even higher levels of expenditures. This regulation is largely designed to control low-cost, low-quality programs offered by academically weak private institutions, particularly at their branch campuses.

Yet assessments of state-granted legitimacy show that the private sector is increasingly becoming accepted as a valid alternative in the provision of higher education. As the data in this study show, almost 3 in 4 private institutions have achieved state accreditation, and a sizeable number of these institutions regularly earn rankings from the Ministry of Education and Science. Russian private universities' enrollments, along with their instructional and supporting facilities, show steady growth. And various cooperative arrangements between nonstate institutions and state universities and other organizations appear to be gaining momentum.

On the whole, nonstate institutions appear to seek legitimacy not so much by conforming to standards and expectations established for state higher education but by developing relationships with influential audiences in the marketplace and by offering valued exchanges to the interested social actors, including state universities and local governmental organizations. They derive their legitimacy not so much from demonstrating their normative and regulatory appropriateness (which they do, as shown by the growing instances of state accreditation) but largely from a logic of their contribution and utility to various stakeholders and constituencies. Their insufficient normative legitimacy may be offset to some extent by their solid base in market-based legitimacy.

Notes

This research was assisted by a fellowship from the International Dissertation Field Research Fellowship Program of the Social Science Research Council with funds provided by the Andrew W. Mellon Foundation. The author also would like to acknowledge support from the Program for Research on Private Higher Education headquartered at the University at Albany-SUNY. The author appreciates comments on the early drafts of this paper from Daniel Levy, Snejana Slantcheva, David Post, and Roger Geiger.

1. The terms *nonstate* and *private* are used interchangeably in this study although they are slightly different in meanings and connotations. The word *nonstate* is generally found in Russian legal discourse and in institutional charters to refer to nongovernmental institutions of higher education that are founded by entities other than the state and independently financed through sources (typically tuition and fees) other than subsidies from the central government. The advantage of using *private* is that the word conveys a number of essential characteristics of these institutions, such as their market origin and entrepreneurial spirit. *Private* is also commonly employed by comparative education scholars worldwide to refer to similar sectors of higher education.
2. Broadly conceived, privatization in Russian higher education is understood as the transference of property rights from government (mainly federal) to other nonprofit or profit-making organizations, that is turning property owned by central, regional, or municipal governments (or the state) into property owned by other entities collectively referred to as nonstate. Nonstate property may be owned by either private or civil (public) organizations, as well as by state-run organizations, such as universities, hospitals, and museums. If, for example, two state-run universities decide to establish a new independent university, this institution will be considered nonstate property (Volkov, Vedernikova, & Rumyantseva 2004).

Fine-grained conceptual descriptions of *private* can be found in Geiger (1986) and in Levy (1986).

3. The word *university* in the plural is used interchangeably with *higher education institutions* to avoid repetition. However, the term *university* implies a high academic status and may not be equivalent to the meaning of Russian higher education institutions. In general, Russian higher education institutions include universities, academies, and institutes, all offering tertiary education degrees consistent with Level 5A programs of the 1997 International Standard Classification of Education.
4. The idea of the degrees of privateness in higher education is explored in more detail in Stetar (1996).
5. The statistics on nonstate enrollments and institutions offered in different sources often vary considerably. For example, Volkov, Vedernikova, and Rumyantseva (2004) provide a figure of 650 institutions currently in operation. This paper uses official sources of statistics provided by the State Committee for Statistics and the Ministry of Education and Science.
6. The expression *state-owned resources* refers to mostly nonmonetary assets owned by federal, regional, and local governments and managed on their behalf by various state-run, mainly nonprofit organizations, including universities, specialized research centers and institutes, hospitals, museums, newspapers, opera houses, etc. These assets include land, physical plant and various facilities, and equipment, as well as government connections as a form of social capital that may be used to provide privileged access to resources (e.g., relief from local taxes or rent). The phrase is not intended to include direct financial transfers from the federal budget as the government generally opposes any form of financing private higher education out of its budget.
7. The study's level of analysis examining the relation between higher education institutions and the environment is an organization field, an intermediate unit between organization and society levels. It delimits the discussion of legitimacy to the influence of actors in a higher education organizational field. As a result, the legitimating effects of broader societal structures and processes, such as public opinion and the mass media, as well as those of intraorganizational stakeholders, such as administrators and faculty, are excluded from the investigation. Also, unlike Clark's *academic guilds*, which stresses the aspects of professionalization, the element of *higher education community* is introduced to emphasize higher education institutions as organizations.
8. In the following sections, the phrase *state universities* refers to both state-run universities and specialized research organizations.
9. This study employs a stakeholder model of organizational legitimacy, emphasizing societal audiences involved in granting legitimacy. It delimits the discussion of the concept of legitimacy to general notions of state-granted and market-based forms. Thorough discussion of the forms of organizational legitimacy, including cognitive (comprehensible and taken for granted), normative (moral or conforming to professional standards of quality), regulatory (legally compliant), and pragmatic (based on actors' self-interest), can be found elsewhere (Ruef and Scott 1998; Scott 2001; Suchman 1995).
10. Accreditation is a comprehensive process of internal and external evaluation of various input, process, and output indicators of institutional activities to

ascertain institutions' conformity to the state educational standards. As an outcome, it means the right to confer academic degrees authorized by the state. Private higher education institutions are eligible to initiate the process of accreditation only after they graduate at least one or more student classes. However, accreditation is not mandatory, and some private institutions continue to operate without this status for years.

11. Clark (1983) distinguishes four sets of values against which institutions of higher education are judged by their relevant audiences: justice, competence, liberty, and loyalty. Preference for *competence* comes in the form of high academic standards, superior qualifications of students and faculty members, emphasis on research and graduate education, and other traditional gauges of academic quality.
12. The relationship between legitimacy and resources along with the relationship between the processes of legitimation and resource acquisition is ambiguous and subject to circular reasoning, and a thorough discussion of how they are related is beyond the scope of this chapter. The view taken here is based on the idea of reciprocal causality: the flow of resources and the construction of legitimacy are mutually reinforcing parallel processes whereby resources are media by which approval and endorsement are expressed (Hybels 1995). In fact, some organizational sociologists view legitimacy as simply one of many resources needed for organizational survival (Dowling and Pfeffer 1975). This circularity can be resolved if one views legitimacy acquisition as a dynamic, continually evolving process whereby resources procured from the environment give rise to increased legitimacy and enhanced legitimacy brings additional resources into organizations (Hybels 1995).
13. Unlike Geiger's (1986) and Levy's (1986) typologies of private higher education institutions that largely take into account the structure of higher education demand or institutional roles and missions, this classification is based on institutional affiliations with founders and owners emphasizing institutions' legitimation orientations and the role of the state, market, and higher education community as legitimizers. Due to its different focus, the present classification is not well suited to account for some institutional types commonly found in the literature on worldwide private higher education (e.g., value-based, religious institutions). Another affiliation-based typology is offered in Bernasconi (2004).
14. The Ministry annually ranks roughly 60 private institutions out of the entire private sector as institutions meeting established standards of academic quality. Institutions that do not provide the requested information are not ranked.
15. This variable is calculated by dividing the area of physical plant in square meters, including instructional and supporting facilities, by a headcount number of students.
16. One must keep in mind that a measure of resource acquisition is also integrated into the evaluation of accreditation, which is granted if higher education institutions are able to demonstrate not only conformity of the content and structure of academic programs to the state educational standards but also an appropriate level of instructional and supporting facilities and faculty members' qualifications (i.e., resources).
17. The third element of the stakeholder framework, the state higher education community, is also a potent source of legitimacy (albeit somewhat ambiguous).

It is part of the state property financed and governed by the central government and also part of the competitive market environment in higher education. Yet it remains an independent, powerful legitimating entity in the higher education organizational field, effecting legitimation through professional authority and norms. As a collective actor in the marketplace, the academic realm also confers pragmatic legitimacy. This realm of influence is discussed in the chapter largely in relation to exchange-based, pragmatic legitimacy derived through partnerships between state and nonstate institutions.

Bibliography

Altbach, P. 1999. Private Higher Education: Themes and Variations in Comparative Perspective. In P. Altbach (ed.), *Private Prometheus: Private Higher Education and Development in the 21st Century*. Chestnut Hill, MA: Center for International Higher Education, School of Education, Boston College, pp. 1–15.

Bernasconi, A. 2004. External Affiliations and Diversity: Chile's Private Universities in International Perspective. Working paper N 4, November 2004. PROPHE Working Paper Series. Retrieved March 15, 2006, from www.albany.edu/~prophe

Clark, B.R. 1983. *The Higher Education System: Academic Organization in Cross-National Perspective*. Los Angeles, CA: University of California Press.

Center for the Monitoring and Statistics of Education (CMSE), Ministry of Education and Science, n.d. Statistika Rossiiskogo Obrazovaniya [Data on Russian Education]. Retrieved March 15, 2006, from http://stat.edu.ru/stat/vis.shtml

Center for Research and Science Statistics (CRSS), the Russian Academy of Sciences 2002. Vissheye obrazovaniye v Rossii: statisticheskii sbornik [Higher Education in Russia: Statistical Annual]. Moscow: CRSS.

DiMaggio, P.J. and Powell, W.W. 1983. The Iron Cage Revisited: Institutional Isomorphism and Collective Rationality in Organizational Fields. *American Sociological Review*, Vol. 48, pp. 147–160.

DiMaggio, P. J. 1988. Interest and Agency in Institutional Theory. In L.G. Zucker (ed.), *Institutional Patterns And Organizations*. Cambridge, MA: Ballinger, pp. 3–22.

Dowling, J. and Pfeffer, J. 1975. Organizational legitimacy: Social Values and Organizational Behavior. *Pacific Sociological Review*, Vol. 18, pp. 122–136.

Etzioni, A. 1987. Entrepreneurship, Adaptation, and Legitimation: A Macro-behavioral Perspective. *Journal of Economic Behavior and Organization*, Vol. 8, pp. 175–189.

Federal Law: *On the Introduction of Changes and Additions to the Law on Education of the Russian Federation 1999*. Moscow: Os-89 Publishers.

Geiger, R.L. 1986. *Private Sectors in Higher Education: Structure, Function, and Change in Eight Countries*. Ann Arbor: The University of Michigan Press.

——— 1991. Private Higher Education. In P.G. Altbach (ed.), *International Higher Education: An Encyclopedia*. New York: Garland Publishing, pp. 233–246.

——— 2004. *Knowledge and Money: Research Universities and the Paradox of the Marketplace*. Stanford: Stanford University Press.

Hybels, R.C. 1995. On Legitimacy, Legitimation, and Organizations: A Critical Review and Integrative Theoretical Model. *Academy of Management Journal*, Special Issue: Best Papers Proceedings, pp. 241–245.

International Finance Corporation 1998. *Investment Opportunities in Private Education in Developing Countries: a Report to the IFC.* University of Manchester, UK: International Finance Corporation.

Kirinyuk, A., Kirsanov, K., and Semchenko, E. 1999. Trudnosti rosta [Difficulties of Growth]. *Visshee obrazovanie v Rosii [Higher Education in Russia]* Vol. 1, pp. 37–40.

Levy, D. C. 1986. *Higher Education and the State in Latin America: Private Challenges to Public Dominance.* Chicago, Illinois: University of Chicago Press.

——— 1992. Private Institutions of Higher Education. In B. Clark and G. Neave (eds.). *The Encyclopedia of Higher Education.* Vol. 2. New York City: New York Pergamon Press, pp. 1183–1195.

——— 2002. Unanticipated Development: Perspectives on Private Higher Education's Emerging Roles. Working paper N 1, April. PROPHE Working Paper Series. Retrieved March 15, 2006, from www.albany.edu/~prophe

——— 2004. New Institutionalism: Mismatches with Private Education's Global Growth. Working paper N 3, January. PROPHE Working Paper Series. Retrieved March 15, 2006, from www.albany.edu/~prophe

Lewis, R.D., Hendel, D., and Demyanchuk, A. 2003. *Private Higher Education in Transition Countries.* Kiev, Ukraine: KM Academia Publishing House.

Lindblom, E.L. 2001. *The Market System: What It is, How It Works, And What to Make of It.* New Haven: Yale University Press.

Ministry of Higher and Professional Education. 2001. O vvedenii pokazatelya ekonomicheskoi ustoichivosti obrazovatel'nogo uchrezhdenia pri litsenzirovanii i akkreditatsii [On the Introduction of the Index of Economic Fitness of Educational Institutions in Licensing and Accreditation Processes]. Retrieved March 15, 2006, from http://www.obrnadzor.gov.ru/oficial_docs/

Neave, G. and van Vught, F. 1994. Government and Higher Education in Developing Nations: A Conceptual Framework. in G. Neave and F.A. van Vught (eds.), *Government and Higher Education Relationships Across Three Continents: The Winds of Change.* Oxford, England: Pergamon Press, pp. 1–21.

Pfeffer, J. and Salancik, G.R. 1978. The *External Control of Organizations: a Resource Dependence Perspective.* New York: Harper and Row Publishers.

Rao, H. 1994. The Social Construction of Reputation: Certification Contests, Legitimation, and the Survival of Organizations in the American Automobile Industry: 1895–1912. *Strategic Management Journal* Vol. 15, pp. 29–44.

Reisz, R.D. 2003. Public Policy for Private Higher Education in Central and Eastern Europe: Conceptual Clarifications, Statistical Evidence, Open Questions. Wittenberg, Germany: Institut fur Hochschulforschung an der Martin-Luther-Universität Halle Wittenberg.

Ruef, M. and Scott, W. R. 1998. A Multidimensional Model of Organizational Legitimacy: Hospital Survival in Changing Institutional Environments. *Administrative Science Quarterly*, Vol. 43, pp. 877–904.

Scott, R. 1998. *Organizations: Rational, Natural, and Open Systems*, 4th edition. Upper Saddle River, NJ: Prentice Hall, Inc.

Scott, W. R. 2001. *Institutions and Organizations*, 2nd edition. Thousand Oaks, CA: Sage Publications.

Solonitsin, V.A. 1998. *Negosudarstvennoe visshee obrazovanie v Rosii [Non-State Higher Education in Russia].* Moscow: Moscow Open Social University Press.

Stetar, J.M. 1996. *Higher Education Innovation and Reform: Ukrainian Private Higher Education, 1991–1996.* Vienna: USIA/USAID.

Stinchcombe, A.L. 1965. Social Structure and Organizations. In J. G. March (ed.), *Handbook of Organizations.* Chicago, IL: Rand McNally, pp. 142–193.

Suchman, M. C. 1995. Managing Legitimacy: Strategic and Institutional Approaches. *Academy of Management Review,* Vol. 20, pp. 571–610.

Tomusk, V. 2002. The War of Institutions, Episode I: Rise and Fall of Private Universities in Eastern Europe. Paper presented at the conference *Advancing Institutional Research Agenda in Education: From Analysis to Policy.* September 19–21, Albany, New York.

Veniaminov, V.N. 2002. *Ot Knyazya Vladimira do Prezidenta Vladimira Vladimirovicha: razmishlenia o negosudarstvennih vuzah [From Prince Vladimir to President Vladimir: Thoughts on Non-State Higher Education Institutions].* Saint Petersburg: International Banking Institute Press.

Volkov, V.V., Vedernikova, E.V., and Rumyantseva, A.E. 2004. *Negosudarstvennii sector visshego obrazovania v Rossii [Non-State Higher Education Sector in Russia].* Saint Petersburg: State University-Higher School of Economics Press.

Williamson, O.E. 1985. *The Economic Institutions of Capitalism.* New York: Free Press.

Chapter Eight

Legitimation of Nonpublic Higher Education in Poland

Julita Jablecka

Introduction

At the turn of the new century, we have witnessed several important world developments that are influencing the shape of higher education today, including formation of a global economy, emergence of knowledge-based societies, increasing internationalization of higher education, spectacular surges of demand for higher education, changed perceptions of the role of education—shifting from a universally available right to an individually secured investment—and the development of nonpublic higher education sectors.

This chapter focuses on issues related to the legitimation of the nonpublic[1] sector of higher education in Poland. The existence and development of any organization is contingent upon its legitimacy, which is taken here to mean a "generalized perception or assumption that the actions of an entity are desirable, proper or appropriate within some socially constructed norms, values, beliefs and definitions" (Suchman 1995, p. 574). The chapter applies the term legitimation to the entire sector of nonpublic higher education, while particular higher education institutions may enjoy different levels of legitimacy.

For analytical purposes, I distinguish between legal and social legitimacy of institutions of higher education. The basis for legal legitimacy can be found in legal higher education acts and provisions, executive orders as well as statutes of particular institutions. Social legitimacy is granted by different stakeholders—the academic community, the local community, applicants and students, and employers. This way, particular stakeholder values and norms are translated into a legal system of social expectations with regard to higher education institutions.

The concepts of legal and social legitimacy are not mutually exclusive. Bearing in mind that they are a social construct, legal norms are created and negotiated in a legislative process and reflect the values and social norms of the participants. Further, the degree and shape of legitimacy may evolve over time. The same pertains to norms, values, and beliefs, that create a foundation or "a benchmark" for action on part of the institutions of higher education themselves. Norms and values of various categories of stakeholders may differ significantly. As a result, an organization that strives to increase its legitimacy among the local community may consequently find it deteriorating among the broader academic community. Finally, even within a given group of stakeholders, the value systems of its particular members (e.g., university rectors versus vocational schools rectors) or its patterns of prioritization might also differ.

This chapter focuses primarily on one particular issue: the influence and attitudes of the public university academic community toward nonpublic institutions. In Poland, representatives of public institutions have played and continue to play an important role in the formation of the legal framework of higher education and in the social legitimation of nonpublic institutions. As in most of Central and Eastern European region, traditional academia has expressed strong negative attitudes toward nonpublic institutions thus contributing to the nonpublic sector's low professional (academic) legitimacy. In this chapter, I present the fast expansion and development of higher education in Poland. The evolution of legal norms over the last 15 years and the role of traditional academia in drafting the legal framework is discussed next. Then the reasons for the low level of academic legitimacy of nonpublic higher education institutions is analyzed. Finally, I conclude with an account of the future implications of these developments for the Polish nonpublic sector.

Evolution of the Polish Higher Education System

In 1990, Poland had 112 higher education institutions. All but one—the Catholic University of Lublin—were run by the state (GUSb).

Before the political change, recruitment rules and admission ceilings were centrally determined. An ideological assumption that the production of intelligentsia exceeded the country's needs helps explain the low admission ceilings to higher education institutions[2]—only 10–12 percent of the age-relevant group were enrolled in higher education. All programs consisted of five-year studies leading to Master degrees. Part-time and evening programs, which also included vocational and engineering studies, were attended by a limited number of adults whose career was hampered by the lack of a higher education degree.

The provisions of the Higher Education Act of September 12, 1990, permitted the development of the nonstate sector of higher education in an attempt to respond to the suddenly rising demand for higher education and to spare the state budget from meeting the whole burden. This trend reflects to a large extent developments across Central and Eastern Europe where "The fiscal incapacity of the state and skyrocketing demand permitted the growth of private higher education institutions" (Galbraith 2003, p. 545). In fact, in Poland, the explosion of demand in higher education by far exceeded prior expectations and resulted from factors such as:

- reversing repressive state control
- changes in the Polish economy[3] stimulating a new need and perception of education as an investment that may yield future benefits
- expectations of higher earnings upon graduation
- unemployment among the less educated
- demographic peak of young people from the age-relevant group (19–24)
- proliferation of secondary schools, while vocational schools—whose graduates were not permitted to continue into higher education institutions—have been gradually reduced (Jabłecka 2006a).

The nonstate sector played an important role in the expansion of higher education in Poland, as it did in many other Central and Eastern European countries. As table 8.1 shows, between academic 1990/1991 and 2004/2005, the number of students increased by 380 percent. In the state sector, the annual rate of growth of the number of students exceeded 10 percent throughout the 1990s and gradually fell to 2.36 percent between academic 2003/2004 and 2004/2005. By contrast, the annual growth rate of students growth in the nonstate sector until the end of the 1990s amounted to several dozen percent (around 80 percent on an average between academic 1993/1994 and 1995/1996[4]), gradually falling, however, most recently to 6.28 percent between academic 2003/2004 and 2004/2005 (GUSc, GUSd). The sudden and rapid growth in enrollment followed by a great slow down fit the regional pattern, but Poland is the starkest examples at least on the nonstate side.

In academic 2004/2005, of the 427 higher education institutions, including those of the Ministry of National Defence and the Ministry of Interior and Administration, 301 were nonpublic, educating 30 percent of the total number of Poland's students (GUSd 2005). Of these, 13 were denominational, run by the Roman Catholic Church and the churches of other denominations. Nonpublic institutions are scarce among certain institutional types and concentrate on others. Of the country's 17 universities,

Table 8.1 Students in state and nonstate higher education 1990/1991–2004/2005, and annual growth

Year	*1990/91*	*1991/92*	*1992/93*	*1993/94*	*1994/95*	*1995/96*	*1996/97*	*1997/98*	*1998/99*	*1999/2000*	*2000/01*	*2001/02*	*2002/03*	*2003/04*	*2004/05*
Students in state h.e.	390 292	413 621	477 445	555 330	634 590	709 431	788 687	871 091	951 116	1 020 318	1 119 201	1 211 379	1 271 728	1 306 225	1 337 051
Annual growth (%)	—	5,98	15,43	16,31	14,27	11,79	11,17	10,45	9,19	7,28	9,69	8,24	4,98	2,71	2,36
Students in non-state h.e.	n.a.	n.a.	16 169	28 937	49 578	89 399	142 928	226 929	331 483	419 167	472 340	509 279	528 820	545 956	580 242
Annual growth (%)	—	—	—	78,97	71,33	80,32	59,88	58,77	46,07	26,45	12,69	7,82	3,84	3,24	6,28

Sources: GUS. 1990–1996. "Szkoły wyższe" (Higher Schools), Warsaw. GUS. 1997–2004. "Szkoły wyższe i ich finanse" (Higher Schools and Their Finances), Warsaw. Calculations by author.

only 1 was nonpublic. Of the 22 technical universities, only 4 were nonpublic, as was just 1 of the 9 agricultural higher education institutions, and 4 of the 22 arts education institutions. But of the 93 economic academies, 88 were nonpublic as were 11 of the 17 pedagogical institutions. Similar tendencies characterized the higher vocational schools[5] where 125 of the 161 institutions were nonpublic (GUSb 2004). The nonpublic sector also enrolls a higher proportion of part-time students than the public sector. While the ratio of part-time students at state institutions has stabilized around 40 percent since 1997–1998, between academic 1994/1995 and 2004/2005, it increased from 53 percent to over 76 percent in nonpublic institutions (GUSc, GUSd). Poland's institutional concentrations, notably public for universities and nonpublic for economic academies, reflect regional and global tendencies.

The future of the nonpublic sector whose growth has been phenomenal has to be considered against the socioeconomic context, which has slowed down the sector recently. The most important factors include the demographics and the job market. Poland faces negative demographic predictions while unemployment persists around 20 percent. Whereas those with higher education find it most easy to find a job, rate of unemployment also has been increased in this group. These factors present enrollment challenges for higher education overall and for the nonpublic sector in particular, especially insofar as it struggles to achieve some of the legitimacy of the public sector.

The Evolution of the Legal Framework as a Source of Legal Legitimacy for the Nonpublic Sector

The regulatory or legal aspect of an organization's legitimacy reflects the extent to which the legitimacy of a particular organizational form depends upon its conformity with explicit rules and regulations (Scott 2001). Although legal provisions have tended to disadvantage the nonpublic sector in Poland throughout its 15 years of existence, they have also dealt with it in ways that avoid strongly undermining or delegitimizing it. On the one hand, the legal framework reflects the nonpublic institutions' weak legitimacy vis-à-vis the public sector. On the other hand, it offers a level of achieved legitimacy, at least enough to meet what Levy describes as a threshold of legitimacy (Levy 2004). At the same time, legal provisions not only reflect but also have impacts, some of which undermine the legitimacy of nonpublic institutions and others promote it.

Three higher education acts, with their respective amendments and executive regulations, have shaped the legal standing of Poland's nonpublic higher education sector over the last 15 years. The creation of the legal framework has been heavily influenced by the public sector academic milieu, both at the stage of the acts' inceptions and at the stage of issuing executive orders by the Minister of Education.[6] Notwithstanding the strong dislike that a significant part of the academic community of the public sector have for the nonpublic sector, the growth of social legitimacy among nonpublic institutions has been such that it has been reflected in the evolving legal norms. A salient aspect of this is the increase in equal treatment of the two sectors, especially in terms of their access to public resources.

The establishment of nonstate institutions of higher education was first made possible with the Act on Higher Education of September 12, 1990 (and the term nonstate was first used). Its provisions enabled the Minister of Education to issue an order concerning the conditions (1) for the establishment of nonstate institutions of higher education, pertaining mostly to the founder's credibility, the financial viability of the institution, its appropriate infrastructure, and so on, and (2) for offering bachelor or master programs, referring to substantive matters and maintenance of quality of teaching contingent on meeting the minimum number of academic teachers with specific academic degrees and titles, employed permanently at a given institution as their first full-time position (in case of studies leading to master degree) or a second one (in case of studies leading to bachelor degree). This executive order also determined the ratio of full-time teachers, holding the title of professor or a degree of doctor *habilitatus* and doctor, to the number of students in different study program groups. These staffing criteria had to be fulfilled both by state and nonstate institutions. The early formal provisions were rather stringent and appeared to be exceptional in the region, where sparse regulation often characterized the early development of private sectors of higher education. The comparatively stiff regulations in Poland reflected a sense that the nonpublic sector could not be trusted without them, that legitimacy required regulations.

Until the establishment of the State Accreditation Commission (PKA), the only form of mandatory control on the part of the state was exercised by the General Council of Higher Education (RGSZW) on behalf of the Minister of Education. Control was exercised in the initial verification of the substantive requirements at the institution's inception. If an application concerning the founding of a new higher education institution failed to meet the criteria set out in these regulations, it was rejected. Resubmission of rejected applications was possible after changes were made. However, neither the stringent criteria nor the rigorous selection and protracted procedure of issuing permits discouraged enterprising founders from persevering in their efforts to set up higher education institutions. This resulted in an explosive development of the nonpublic sector.[7]

Nonstate institutions viewed the 1990 Act as largely granting them substantial autonomy (see Pawłowski 2004). Some winced, however, at what they have seen as insurance of unequal competition between the two sectors on the educational market. State institutions have continued to receive state subsidies for their didactic activities, research, and bursaries. Although the Act did not rule out state support for the nonstate institutions, such funding would have required issuance of special executive orders. At least 90 percent of nonstate institutions' revenue still consists of tuition fees, leaving the institutions vulnerable in this market dependency.

Subsequent regulations under the same Act allowed state institutions to collect tuition fees from part-time students and those who attended evening courses or applied only to appear for exams. On the one hand, the regulations have at least limited the nonstate sector's privilege of state subsidy but, on the other hand, they have allowed the state institutions to compete for private money—thus undermining an otherwise distinctive advantage that nonstate institutions enjoyed. The combination of this access with a continuation of privileged state subsidy has contributed to a nonstate view of unequal intersectoral treatment, making it harder for the nonstate sector to achieve its goals, including enhanced legitimacy.

The Act on Higher Vocational Schools was passed in 1997. It was initiated by the staff of the Ministry with the intention to tune the educational system to the needs of the labor market. The vocational higher education institutions that appeared after this Act were to have a more practice-oriented profile: they could employ lecturers with practical, for example, industry experience, offer students vocational programs, and students themselves were required to undergo a 15-week vocational apprenticeship. Thenceforth, new nonstate institutions of higher education have been established based only on this Act. Institutions that met this criteria received the status of vocational institutions. The Act also introduced a Higher Vocational Schools Accreditation Commission that has been received unfavorably by both the majority of state institutions, which were concerned that the level of education would deteriorate, bearing in mind that vocational schools do not have the obligation to carry out research or emphasize faculty development, and the nonstate institutions, which were concerned by the heavier teaching workloads imposed by the new state and nonstate vocational institutions as well as by the fact that vocational schools could offer only bachelor degrees and, even if meeting appropriate requirements, could not apply for permission to offer master or doctoral courses (which these nonstate institutions established before the 1990 Act could do). At the same time, the Act threatened the nonstate sector by facilitating the establishment of state vocational institutions away from traditional academic centers, in the periphery of the country, where in mid-1990s nonstate institutions enjoyed a monopoly.

At the end of the 1990s and in the early 2000s, further legal decisions continued to influence the development of the nonstate sector. The first of them was an amendment to the Acts of 1990 and 1997, which established the PKA. Since 2002, both state and nonstate higher education institutions are to undergo accreditation following the same criteria. Successive new regulations important for the nonpublic sector have been initiated, as in the case of the Act on Vocational Higher Education Institutions, by the Ministry and in response to increasing reproaches that the existing legal system failed to observe the constitutional principle of equal treatment of all citizens. As a result, new regulations were introduced entitling full-time students at nonstate institutions to financial support at the same conditions as at state institutions; the same regulation was soon extended to part-time students as well. Executive regulations also made it possible for nonstate institutions to obtain subsidies from the state budget for didactic activities and for capital projects (albeit under so stringent conditions that only a few institutions managed to qualify). Nonstate higher education institutions (except vocational ones) were also allowed, on similar conditions as state ones, to apply for grants and statutory subsidies; competition terms were, however, very rigorous. These regulations aimed at reducing differential treatment of the two sectors. Crucially, they also reflect the rising legitimacy of the nonstate sector in the eyes of state authorities and the academic community. The conditions for obtaining such state assistance, however, left most nonstate institutions without realistic prospects. As a result, the nonstate sector community continued to perceive the attitude of the state and the academic community as hostile.

It should be noted that the state authorities responsible for education after 1990, consisting mostly of liberal-minded representatives of the pre-1989 Polish political opposition among the academic community who were intent on introducing competition in education,[8] played a strong role in shaping the first act on higher education (as far as the nonstate higher education sector was concerned). In the following decade, the creation and implementation of the legal higher education framework has continued to be dominated by the academic community of the state higher education institutions (who exercise an advisory function to the Minister of Education). At the system level, this group has been represented by the RGSZW composed of elected representatives of the state higher education institutions (a representative of the nonstate higher education institutions participated in the Council's work as an observer). The General Council not only drafted the staffing criteria and—up to the end of 1990s—the so-called teaching standards for the minister but (as I mentioned earlier) also assessed applications for the establishment of nonstate higher education institution or the launching of study programs. Based on these assessments, the minister issued or rejected institutional permits.

Recently, the General Council's duties have been passed on to the PKA, whose main function is to evaluate the teaching process and program compliance with specific requirements. The PKA is composed of representatives of higher education institutions (in practice almost all from state ones) and the Student Parliament. No official representative of the nonstate higher education sector has been included—again reflecting this sector's weakness in legitimacy. During its first three-year term, the PKA evaluated 105 state and 132 nonstate institutions of academic (operating according to the Act of 1990) and vocational status (the Act of 1997). Of these, the state higher education institutions operating according to the Act of 1990 obtained 79 percent of the positive and excellent grades and nonstate institutions—67 percent of such grades. State institutions operating according to the 1997 Act obtained 82 percent of such grades and the nonstate—71 percent (PKA 2004). These figures speak of the strongly negative assessment of the quality of the nonstate sector on the part of the state sector, notwithstanding whether assessment is valid or prejudicial.

Although legal acts failed to acknowledge the role of the conferences of rectors as formal opinion bodies advising the Minister of Education, the Conference of Rectors of Polish Academic Schools (KRASP) exerted substantial influence on the process of shaping the regulations, while the role of the Conference of Rectors of State Higher Vocational Schools and the Conference of Rectors and Founders of Nonstate Higher Education Schools was quite limited.

Since the mid-1990s, faculty members of state academic institutions were heavily involved in the three successive teams drafting a new Act on Higher Education. A representative of the nonstate higher education institutions participated only in the third team, officially appointed by the Polish president. All teams have been dominated by the rectors of the most prestigious state academic institutions. The drafting of the future act was accompanied by strong protests on part of the representatives of the nonstate higher education against the lack of consultation with them; they also expressed the view that some of the proposals ignored the interests of the nonstate sector. Hot debates at universities, in the press and on Internet also took place both during the drafting of the act and the parliamentary discussions on it. In addition, the Solidarność Trade Unions submitted their own draft in the form of a member of parliament's motion. A controversial topic in these debates, deeply concerning the nonstate sector, focused mostly on the employment of faculty on multiple full-time positions at the same time. Had a proposal to allow for a single full-time position for faculty members been passed, it would have implied the collapse of a significant number of nonstate higher education institutions.

The final version of the Act, passed in July 2005, is, however, rather favorable to the nonstate sector. With respect to employment, faculty may be employed at no more than two full-time positions, regardless of sector.[9]

Breaching this may be regarded as reason for dismissal from the institution that is their primary employer. Taking a second full-time position without the approval of the rector of the primary employer, or without notifying the rector, also carries the risk of dismissal. No restrictions have been imposed, however, on employees with only part-time contracts.[10] Although these provisions on faculty apply to both sectors, they affect them very differently. Nonpublic institutions rely mostly on part-time faculty. Thus, these institutions can be hurt by restrictions on faculty from public universities teaching in multiple nonpublic institutions or even obtaining a full-time post in a nonpublic institution. On the other hand, nonpublic institutions have ample flexibility in how they hire their own part-time faculty or employ public university full-time faculty who do not work part-time in other nonpublic institutions. Public universities, in turn, are protected from their full-time faculty devoting much of their time to multiple nonstate institutions but may suffer from the inability of their full-time faculty to make adequate supplementary income. These public-private dynamics are common in the region and beyond and again fit an intersectoral legitimacy gap.

The new Act also specifies requirements regarding the founding of both nonpublic and public institutions of higher education, procedures for closing of institutions in both sectors, and dismissal of their rectors. Ministerial permits to establish a nonpublic higher education institution are valid for five years, though extension may be denied should the institution's study programs receive negative evaluations from the accreditation committee or should the institution or its founding body be judged to be in major infringement of the laws.

The earlier distinction between the higher education institutions established under the 1990 Act and the 1997 Act on Higher Vocational Schools was replaced by a distinction between academic and vocational institutions thus officially introducing stratification of institutions of higher education.[11] The new Act also embraced a series of earlier legal arrangements, which confirmed the equalization of the two sectors with respect to conditions for receiving state funding and employment regulations. One important new regulation, from the point of view of equalization of the competing terms, stipulates that no support is to be provided for public higher education institutions toward teaching fee-paying students (those studying in part-time and evening courses).

Nonpublic institutions of academic status have legally authorized representation in the KRASP while nonpublic vocational institutions are represented in the KRWSZ. At present these bodies encompass both the public and the nonpublic sectors. On the other hand, the Conference of Founders and Rectors of Nonpublic Higher Education Institution (KZRNWSZ) which has been in existence for over ten years, failed to be acknowledged by the Act as a separate, official representation of nonstate institutions of higher education.

In sum, the new Act lays out rather detailed requirements for all institutions of higher education. With respect to the nonpublic sector, it reflects and pushes forward the evolution of the legal framework as an important source of legitimacy, though not by any means equalizing the legitimacy of the two sectors. Noticeable also is the gradual convergence of the legal conditions in which both the public and nonpublic higher education institutions operate in Poland, notwithstanding the differential intersectoral impacts of these legal conditions.

Explaining the Weak Professional Legitimacy of Nonpublic Institutions

"Academic/professional/legitimacy," as Reisz notes, "is a social concept. It gains its meaning from the social acceptance of an institution as 'academic' . . . refers to expectations related to the 'academic-ness' of an institution" (Reisz 2003, p. 24). Whereas legal legitimacy has increased for nonpublic institutions, the picture is less favorable regarding their professional legitimacy. As mentioned earlier, negative attitudes toward nonpublic institutions on part of the academic community from public institutions accompanied the formation of the legal higher education framework pertaining to the nonpublic sector. The reasons for these attitudes are complex and relate primarily to the low level of compliance of nonpublic institutions to dominating norms and academic values in the public sector. According to the majority of members of public institutions, nonpublic institutions fail to meet traditional requirements with regard to higher education.

Other reasons—some interrelated—for the lack of nonpublic professional legitimacy include the following.

Offering Predominantly Bachelor Programs

Poland's institutions of higher education were historically shaped by offering the traditional model of uniform studies leading to a master degree. The 1990 Act divided this traditional degree into two tiers: bachelor and master. It is the former that dominates at nonstate higher education institutions. The 1997 Act on the Higher Vocational Schools provided for bachelor programs only. Most of Poland's nonpublic institutions have been established according to the Act of 1997. Moreover, any nonpublic institution (including also those set up under the Act of 1990, which could offer master and doctoral programs) intent on launching a new study program, had to apply first for a permit to launch bachelor level courses only. More liberal provisions for launching bachelor than master programs certainly help the nonpublic sector in the higher education marketplace which in turn, allows it to build its legitimacy here. But the distinction between bachelor and master highlights the nonpublic sector's low standing in professional legitimacy. Under the present Act, which bestowed the status of an academic higher education institution only to those institutions that were entitled to confer doctoral degrees in at least one discipline, almost all nonpublic institutions will become vocational institutions since only about 25 percent of them are currently entitled to offer master programs, only 6 of them can confer doctoral degrees, and just a single one has the right to award the second doctorate (doctor *habilitatus*). Seen from the leading academic institutions, nonpublic institutions are inferior type of institutions.[12]

Student Qualifications

While public institutions admitted students to their daily courses through competitive exams (there were no admission exams for applicants for part-time and evening studies), applicants to nonpublic institutions did not have to pass admission exams. The new secondary school national graduation exam, which replaces institutional admission exams for (both sectors) higher education, was introduced in Poland only in 2005. Under the new Act, candidates studies may be selected based on the secondary school graduation exam's results only and the introduction of any additional admission procedures needs to be approved by the Minister of Education. However, this does not imply that all those who have passed the exam have to be admitted, for selection may also be determined by a given required number of points obtained on the graduation exam. Applicants can apply simultaneously to several study programs and choose between programs upon admission. The free full-time studies (to avoid distorting the ratio of students in particular type of courses, according to state regulations, state institutions have to teach at least 50 percent of their students for free) at prestigious public institutions come first in this choice, while paid studies at both state (fees may be charged for evening and extra-mural studies as well as for external exams) or nonstate institutions are considered usually a second or a third choice. Free daily studies are most popular with applicants whose number often exceeds several times that of the available places. Thus, the academically strongest applicants are directed toward state institutions while the educational preparation of students who undertake their studies in nonpublic higher education institutions is lower. This sort of preferential ranking is common in systems that have a private sector and both fee-paying and subsidized students in the public sector.

Low Quality of Education and Predominance of Part-Time and Evening Studies

Representatives of state higher education institutions upbraid the low level of teaching at nonpublic institutions. An assessment of the PKA, quoted earlier, partially supported this criticism although the differences in quality between the two sectors were not as substantial as the opinions circulating in the milieu claimed. It is, however, a different aspect of the operation of nonpublic higher education institutions, which the PKA fails to take into account, namely a chance for the student to find a job and to succeed in the labor market. The most popular fields of study in nonpublic institutions established under the Act of 1990 include management and marketing, economics, pedagogics, and administration. Studies by Szanderska et al. (2005) point out that similar programs dominate also at nonpublic vocational schools. However, it is these studies, that produce most unemployed graduates. As a result, many nonpublic institutions of higher education contribute to structural unemployment (Trzeciak 1998, KUP 2002), or at least are perceived to do so, which undermines their legitimacy. Research in nonpublic institutions has lead to the conclusion (albeit an intuitive one since no surveys of the candidates' expectations have been carried out) that institutions do not correlate their founding, the launching of new study programs, and the admission quotas to labor market demands but instead they cater to student demand (Sztanderska et al. 2005). In this context, one might conclude that because of negative demographic trends and growing competition for students, higher education institutions will have to adapt themselves to student and labor market expectations.

Another factor that corroborates the low opinion of the quality of education in nonpublic institutions is the predominance of part-time and evening studies over full-time studies. Part-time studies are generally regarded by the academic community as inferior, run for students who are often in employment and commute once a month to classes held on weekends. While they last as long as daily classes, students attend fewer classes. It is under this regime, however, that the majority of the students in the nonpublic sector are taught.

Small Institutional Size

Most nonpublic institutions are often small, sometimes with no more than a few hundred students and with just a single study program. Such institutions have difficulties maintaining strong academic atmosphere.

Distant Location

Located away from established academic centers, many nonpublic institutions lack the critical mass necessary to form an academic community and are detached from established information and communication resources.

Little Research

Although eligible for research funding, most nonpublic higher education institutions conduct very little scientific research. Out of over 300 institutions, only a few applied for research funding in 2005 and received less than 1 percent of all funds received by the higher education sector. In addition, most of the conducted research is applied rather than basic, often either commissioned by the business sector or in cooperation with it. And basic research is generally valued higher than applied by traditional academic institutions.

Faculty Profile and Commitment

Since faculty development and training takes time, the new nonpublic higher education institutions were compelled to resort to staff who already had been employed elsewhere, to take them over, or to employ them part time. As already mentioned, in order to comply with minimum staffing requirements necessary for an institution to offer bachelor or master programs, the institutions have to employ on full-time basis a determined number of scholars with appropriate qualifications. To obtain academic staff for a permanent but the only full-time position is very difficult for the private higher education institutions bearing in mind that a professor at a state higher education institution has secure employment, the state institutions enjoy greater prestige than the nonstate ones (with very few exceptions), generate academic atmosphere, and have academic community enabling scientific development. The lack of academic self-sufficiency, faculty members holding multiple positions as well as the fact, pointed out by Poland's equivalent to National Audit Office–NIK that a significant member of institutions fails to meet the staffing minimum, (Forum Akademickie 2000) is another delegitimizing force.

Nonstate institutions are often located in the provinces and their staff remains outside the mainstream of scientific life. Finally, an overwhelming majority of private higher education institution do not carry out any research at an European level, perhaps only those of a regional nature. Of these institutions, 99 percent do not obtain any funding from the state budget for research while chances to obtain an individual research grant by a member of faculty of a nonstate institution lacking credentials and a scientific community are scant. One compounded result is that often the faculty members at nonpublic institutions may be easily promoted and obtain professor's positions with little academic achievements. Finally, faculty at nonpublic institutions, particularly of those that are located away from large urban agglomerations, are perceived as persons who having conducted their classes leave the town and are interested neither in their students nor in conducting research.

Commercialization of Educational Services

The emergence of nonpublic higher education institutions (as well as the introduction of paid studies at public institutions) signaled the appearance of internal competition and the commercialization of educational services. To regard education as merchandise contradicts the value system of most of the traditional academic community for whom education is a public good.

The Employment of Advertising

Dishonest competition through using the brand and the faculty of a public higher education institution to advertise "piggy-back" a nonpublic one has often been employed. A significant part of the academic community regards such practices on part of the nonpublic institutions as theft. Such attitudes are also shared by these nonpublic institutions that enjoy considerable prestige.

Governance System

The governance of Polish higher education institutions is based on the strong role of collegiate bodies that have decision-making powers and a relative weak role of those organs that operate through individual office holders. Most nonpublic institutions, on the other hand, have developed a managerial system under which the role of collegiate bodies is limited. The academic milieu perceives this as a restraint on traditional forms of self-governance of the scholarly community.

In response to many of these criticisms, some nonpublic institutions of higher education have attempted to adapt to accepted professional standards and expectations. Different multidimensional strategies have been used to increase their academic legitimacy. Thus some institutions apply for the right to confer master (and even doctoral) degrees, which also implies an appropriate faculty recruitment strategy. Increasingly, nonpublic institutions hire the staff employed at research institutes as well as the retired staff of prestigious universities. Another strategy consists of extending the profile by launching new study programs currently in demand. Partnerships with foreign institutions have also been common in the nonpublic sector, through which institutions have conferred joint degrees together with a foreign institution and exchanged their faculty and students. The strongest nonpublic higher education institutions applied for their accreditation to one of the voluntary accreditation commissions acting under the aegis of the KRASP or of an independent body, SEMI-FORUM, Association of Management Education, established as the first voluntary accreditation body in Poland in 1993. The SEMI-FORUM Association confers voluntary accreditation and sets a high threshold. Obtaining such an accreditation considerably raises institutional prestige (Ratajczak 2004). Public–nonpublic partnerships also emerged where institutions that do not offer master programs signed agreements with public institutions to have their bachelors admitted to the latter without entrance exams. A significant part of the nonpublic higher education institutions also began to offer postgraduate studies that are not, however, subject to accreditation and are difficult to evaluate. Finally, some nonpublic higher education institutions capitalized on their unique character, organizational culture, and academic climate. They developed their library resources based on modern media, introduced alternative teaching methods, established their own system of academic scholarships, grants and bursaries for students and started recruiting foreign students.

Regional and local communities as a source of legitimacy

The last 15 years have witnessed gradual legal equalization of the public and nonpublic sectors and the growing legal legitimacy of the nonpublic institutions. Despite difficulties in raising their legitimacy among the academic community, nonpublic institutions have advanced in legitimacy among regional and local communities, and among the regional authorities in the provinces. Initially, nonpublic institutions were established mostly in close proximity to large urban agglomerations that offered an easily accessible pool of faculty employed by state higher education institutions and available resources. As a result, for example, Warsaw has 77 higher education institutions of which 62 are nonpublic. Later, nonpublic institutions gradually started appearing away from large agglomerations. Although most little towns and communities would like to have their own institution of higher education, it takes time for this community to appreciate the advantages of having a nonpublic one. This initial lack of trust in the institution often results from distrust toward its founders, who are unknown in the local community, and an apprehension that it is only the founder who will benefit from the establishment of a nonpublic school. However, both the authorities and the local community often begin to appreciate the advantages in the availability of an educational institution in the near proximity such as stronger business activity and enhanced community prestige, increased opportunities for participation in higher education and for combining work and study, and increased incentives to remain in the local community after graduation (Stachowski 2000).

Studies carried out in nonpublic higher education institutions located in Poland's eastern territory indicated that often close to 40 percent of students are of rural background, compared to slightly under 30 percent in central Poland, which exceeds the relevant percentage among students of academic higher education institutions (Stachowski 2000, Kruszewski 2000). Institutions of higher education contribute to regional development. They provide direct employment and stimulate capital projects such as construction of infrastructure for the institutions, hotels, pubs, restaurants, cafés, and social facilities (Jabłecka 2006b). Thus, the future of the majority of nonpublic institutions located away from large agglomeration centers could be seen in their local orientation.

Final Remarks

The evolution of regulations provides evidence of the gradually rising level of legal legitimacy of nonpublic higher education institutions, expressed in

- the gradual introduction of an operating framework and formal requirements uniform for both public and nonpublic sectors
- the enhancement of the academic character of standards for both sectors and the introduction of hierarchy (stratification) of higher education institutions based on those standards
- the gradual convergence of the terms of competition on the educational market for both sectors.

The analysis reveals that the low degree of academic legitimacy of nonpublic higher education institutions among the state academic sector, on the other hand, is rooted mostly in a difference of standards, hierarchy of values, and of missions rather than in certain formal irregularities or unethical competition. Poland is a typical European example of a country where the academic community is accustomed to a uniform system of education consisting of higher education institutions that operate under similar standards and carry out a similar mission, and the Humboldtian emphasis on the symbiosis between science and teaching. To this community, a system of education with differentiated levels and objectives is alien.

Three different strategic approaches are available to nonpublic higher education institutions in their future attempts to enhance their legitimacy among different social groups:

- a strategy of emulation of academic higher education institutions based on deliberate adoption of mimetic conduct (DiMaggio and Powell 1983). This strategy has been already applied by a handful of higher education institutions that have been quite successful in imitating features and standards of the academic sector
- a strategy of adaptation to regional demands
- a strategy of specialization that could be adopted by those higher education institutions that do not belong to any of the previous groups, and consists of offering specific or unique studies or enter in partnerships with other nonpublic institutions or with higher education institutions abroad.

Notes

1. According to the Act on Higher Education of September 12, 1990, the term "non-state higher education" was used in legal documents in Poland rather than "private higher education" since founders of the nonstate higher schools were natural persons or legal persons: limited liability companies, associations, cooperative societies, and foundations. A term "nonpublic higher education" was introduced under the new Act of Higher Education of July, 2005. This term was also used in the new Polish Constitution of April 2, 1997. The nonpublic higher education institutions have the nonprofit status and are bound to reinvest the whole profit in their development. In the chapter, the terms public and state are used interchangeably.
2. Growth of the number of students in higher education was centrally repressed by imposed tight admission ceilings.
3. The developments within the Polish economy stimulated the aspirations of young people and their parents for higher education related to expectations for better positioning on the labor market upon graduation (than people with no tertiary education).
4. Bearing in mind that new nonpublic institutions were being established in those years.
5. Higher vocational schools began to develop since 1998 following the 1997 Act on Higher Vocational Schools.
6. The minister responsible for education took into account the General Council on Higher Education—the Polish representation of the state academic self- governing body.
7. Stringent application of those regulations can be illustrated with statistical data for selected years on the number of applications for a founding permit submitted

to the Minister of Education (for the first time) and those that have been rejected: 6 applications were submitted in 1991, 5 of which were rejected. In 1994, 35 new applications were submitted of which 14 were rejected. In 1998, 38 new applications were submitted of which 14 were rejected (State Accreditation Commission 2004). It should be borne in mind, however, that the regulations referred to concerned the entry conditions at the moment of the establishment of a higher education institution. How well those requirements were observed on an on-going basis cannot, however, be attested since the State Accreditation Commission did not come into existence until 2002.

8. These included the minister of education, the deputy-ministers of national education and even department directors.
9. For bachelor degree programs, faculty members that are employed in their second full-time job can also be counted toward the required staffing minimum. However, the required staffing minimum for an institution allowing it to confer a master degree demands a member of faculty employed there on a full-time permanent position.
10. One habilitated faculty member can be counted toward fulfilling the single staffing minimum in master programs at their primary institutions or toward two staffing minima in bachelor programs. Faculty student ratios in a given program are determined by the minister.
11. Criteria of the past and the present classifications of institutions of higher education are different. For instance, nonstate higher vocational schools set up under the Act of 1997 were distinct from those founded under the Act of 1990 in three aspects described earlier: at vocational schools, the profile of teaching within study programs or specializations was more vocational and practical; they were under no obligation to carry out research and develop their faculty; they could also offer only bachelor degrees. At present, vocational and academic higher education institutions are differentiated by the degrees they are authorized to confer. The vocational, specialized teaching profile is not mentioned in the regulations; and they may confer master degrees as well; a higher vocational education institution may also become an academic one by obtaining the right to confer a doctoral degree in at least one discipline. Vocational higher education institutions continue not to be bound to carry out research or to develop their faculty professionally. All higher education institutions, on the other hand, have been hierarchized. The top level in the hierarchy of academic sector is occupied by universities and technical universities, which, in order to be able to be referred to as such, need to be entitled to confer a doctoral degree in at least 12 different fields. Beneath them, there are polytechnics (in four fields) and academies (in two). Vocational institutions occupy the bottom level. This hierarchy as well as the ability of vocational institutions to offer master degrees reduced the prestige and the legitimacy of nonpublic institutions as a whole. As a result of the new regulations, most nonpublic higher education institutions automatically obtained the status of vocational institutions, or those at the bottom of the professional hierarchy of academic institutions.
12. Similar opinions have been voiced also with regard to state higher vocational schools.

Bibliography

DiMaggio, P. and Powell, W. 1983. The Iron Cage Revisited: Institutional Isomorphism and Collective Rationality in Organizational Fields. *American Sociological Review*, Vol. 48, pp. 147–160.

Galbraith, K. 2003. Towards Quality Private Higher Education in Central and Eastern Europe. *Higher Education in Europe*, Vol. 28, No. 4, December, pp. 539–559.

GUS a [Main Statistical Office]. 1998. *Losy zawodowe absolwentow* [*Careers of Graduates*]. Warsaw.

——— b. 1991–2004. *Roczniki statystyczne* [*Statistical Yearbook*]. Warsaw.

——— c. 1990–1996. *Szkoly wyższe* [*Higher Schools*]. Warsaw.

——— d. 1997–2004. *Szkoły wyzsze i ich finanse* [*Higher Schools and Their Finances*]. Warsaw.

Forum Akademickie. 2000. No. 7–8. Kontrola NIK w MEN, Niepaństwowe na cenzurowanym [National Audit Office's investigates the Ministry of National Education, Non-State HEIs under Scrutiny].

Jabłecka J. 2006a. Nonpublic Higher Education in Poland. *Higher Education in Europe*, in preparation.

Jabłecka, J. 2006 b. Regional role of tertiary education. In Dabrowa Szefler, M. and Jabłecka, J. *OECD Thematic Review of Tertiary Education, Country Background Report for Poland*. At www.oecd.org/dataoecd/49/55/37231744.pdf, last accessed October 20, 2006.

KUP. 2002. *Bezrobotni absolwenci szkół ponadpodstawowych wg zawodów wyuczonych w II półroczu 2001* [*Unemployed Graduates of Secondary Schools by Vocations Learned In 2H2001*]. Warsaw: Krajowy Urząd Pracy Publisher.

Kruszewski, Z. 2000. Case study 3: Szkoła wyższa im. P. Włodkowica w Płocku [Case Study 3: P. Włodkowic's HEI in Płock]. In Misztal B. (ed.), *Prywatyzacja szkolnictwa wyższego w Polsce* [*Privatisation of Higher Education in Poland*]. Cracow: Universitas.

Levy, D.C. 2004. The New Institutionalism: Mismatches with Private Higher Education Global Growth. A PROPHE Working Paper # 3 at http://www.albany.edu/dept/eaps/prophe/publication/InstitutionalismWP3.htm, last accessed October 4, 2006.

Misztal, B. (ed.) 2000. *Prywatyzacja szkolnictwa wyższego w Polsce* [*Privatisation of Higher Education in Poland*]. Cracow: Universitas.

PKA. 2004. *Sprawozdanie z działalnosci w I kadencji 2001–2004* [*Report from Activities in the First Term of Office, 2001–1004*]. Warsaw: Panstwowa Komisja Akredytacyjna.

Pawłowski, K. 2004. *Społeczeństwo wiedzy, szansa dla Polski* [*The Knowledge Society – an Opportunity for Poland*]. Cracow: Znak.

Ratajczak, M. 2004. Środowiskowa i państwowa akredytacja szkolnictwa wyższego [Community and the State-led Accreditation of Higher Education Institutions]. *Nowe Zycie Gospodarcze*, June 20.

RGSW. Different years. Sprawozdania z działalnosci za lata 1991–2000 [Reports from Activities, 1991–2000]. Warsaw: Rada Główna Szkolnictwa Wyższego.

Reisz, R.D. 2003. Public policy for private higher education in Central and Eastern Europe. Conceptual clarifications, statistical evidence, open questions, HoF

Arbeitsberichte 2/2003, Wittenberg: Institut für Hochschulforschung, Martin Luther Universität Halle Wittenberg (Institute for Higher Education Research at the Martin Luther University Halle Wittenberg).

Scott, R. 2001. *Institutions and Organizations*. London: Sage Publications.

Stachowski, Z. 2000. Case study 2: Projekt tyczyński [Case study 2: the Tyczyn Project]. In Misztal, B. (ed.), *Prywatyzacja szkolnictwa wyższego w Polsce* [*Privatisation of Higher Education in Poland*]. Cracow: Universitas.

Szefler, M. and Jabłecka, J. 2006. *OECD Thematic Review of Tertiary Education, Country Background Report for Poland.*

Sztanderska, U., Minkiewicz, B., and Bąba, M. 2005. *Oferta Szkolnictwa Wyższego a wymagania rynku pracy* [*The Offer on the Part of Higher Education and the Requirements of the Labour Market*]. Warsaw: Krajowa Izba Gospodarcza and Instytut Społeczeństwa Wiedzy.

Suchman, M. 1995. Managing Legitimacy: Strategic and Institutional Approaches. *Academy of Management Review*, Vol. 20, No. 3, pp. 571–610.

Trzeciak, W. 1998. Analiza rynku pracy absolwentów szkół wyższych [Analysis of the Labour Market of the Higher Education Graduates]. In Buchner Jeziorska, A., Minkiewicz, B., Osterczuk A. (eds.), *Studia wyższe, szansa na sukces* [*Higher Studies – an Opportunity of Success*]. Warsaw: Instytut Spraw Publicznych.

Chapter Nine

Institutional Efforts for Legislative Recognition and Market Acceptance: Romanian Private Higher Education

Luminiţa Nicolescu

Introduction

Recent changes in higher education systems worldwide have stimulated the emergence of new forms of higher education. (Kovac et al. 2003; Maasen and Stensaker 2003; Mora and Vila 2003). As Romania's new institutions multiply at a high speed, private institutions in particular have confronted the issue of legitimacy (Nicolescu 2003b; Brătianu 2002).

The main argument of this chapter is that the legitimacy of private higher education in Romania is closely linked with the public sector. For private institutions of higher education and legislators, public universities serve as a key benchmark in assessing quality of teaching and learning. Since 1990, public higher education has been recognized by all constituencies as *the* standard in Romanian higher education. In addition, the legitimacy of private higher education in Romania has contradictory characteristics that have evolved unevenly over time due to changes in the environment:

1. Different constituencies evaluate private higher education differently. Such constituencies include the public at large, the legislative organs, the market and academia. However, at the same time, most of these variations of evaluation fit into a general image of private higher education.
2. Although the legitimacy of private higher education institutions has been continually increasing, they still bear the illegitimacy stain.

This chapter begins with a definition of legitimacy. It then looks at the development of private higher education in Romania after the political changes of 1989. Next, the chapter offers a perspective on four different components of the legitimacy of the private institutions in Romania.

Different Components of Legitimacy

For the purposes of this analysis, I employ a broad definition of legitimacy of higher education. Legitimacy is seen here as the recognition, credibility, and acceptance given to a higher education sector in general (and to an institution in particular) by different constituencies, (1) the public at large, the society, (2) the legislators, (3) the direct beneficiaries of higher education (namely students, graduates, and employers), and (4) the academic community.

Legitimacy of higher education institutions can be approached through a consideration of the *role* that a higher education sector is perceived to play in the society at large. There are opinions (Altbach 1999) that a shift in the perception of the role of higher education exists from it being a public good that contributes significantly to society by imparting knowledge and skills to those who it educates, to it being more of a private good benefiting the individual more than the society as a whole. In this context, one can assume that the larger the contribution of higher education to both society and individual in general, and of private higher education in particular, is perceived to be, the higher the acceptance of the sector and its legitimacy.

The degree of legitimacy can also reflect the compliance of a sector or an institution to given standards and norms. Thus, the legitimacy of higher education, including private higher education, might increase as institutions adopt standards and norms required by different evaluating organizations, either NGO's, governmental, or independent academic organizations. In their attempt to increase legitimacy, institutions direct their efforts toward the acceptance of different stakeholders by either setting goals and institutional practices that are credible or acceptable, or setting new goals and institutional practices for which to start to build credibility. Institutional practices include aspects such as managerial techniques, developing curricula, teaching techniques, academic staff and its development, communication flows—student/professor, student/administration, support services, and technological change. Institutional efforts to increase legitimacy encompass the conscious creation of a positive image about the role of private higher education in a society or trying to comply with the known expectations of the public.

Hence, four major aspects of legitimacy are useful to this present study of the private sector in Romania. Figure 9.1 presents the framework used to

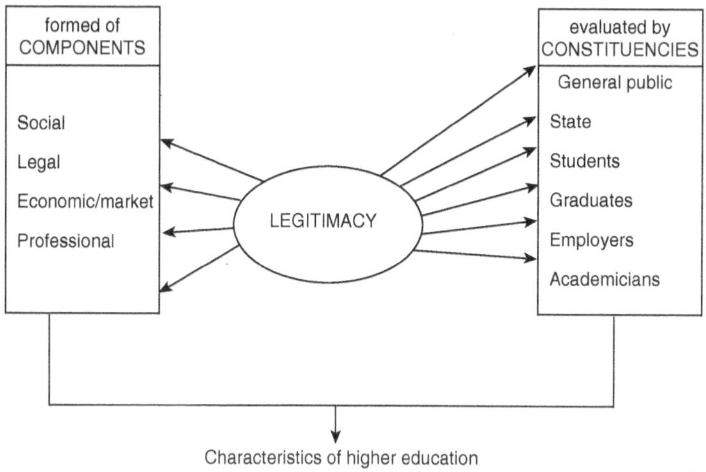

Figure 9.1 Framework of legitimacy analysis

discuss the legitimacy of Romanian private higher education:

- social legitimacy (or moral legitimacy) that looks at the acceptance of private higher education by the general public, based on the perceived roles of private institutions;
- legal legitimacy that refers both to the legal acceptance of the private higher education sector and to the level of compliance of private institutions with legal requirements—expressed through accreditation in the Romanian case;
- market legitimacy that reflects the cost/benefit analysis of higher education services for its main beneficiaries: students, graduates, and employers;
- professional legitimacy (or academic legitimacy) that considers the degree of compliance with academic standards, referring to the acceptance by the academic community.

Development of the Romanian Private Higher Education

Romania is one of the Central and Eastern European countries that experienced an unprecedented growth and transformation of its higher education system after the fall of communism. From its very beginning in the early 1990s, private higher education developed rather as a chaotic process than as a result of an organized reform.

The development took place in the context of a dramatic expansion of the higher education sector in response to the excess demand, induced primarily by the over-restricted number of students admitted to tertiary education prior to 1990. Similar to Latin America (Geiger 1988, Levy 1986) and other Central and Eastern European countries (Kwiek 2003, Tomusk 2003), private initiative in Romania quickly responded to the excess demand. The largest development of the private sector took place in the first years after the opening of the Romanian economy and was reflected in:

1. Increase in the number of private higher education institutions: In the period 1989–2004, the total number of higher education institutions in Romania multiplied almost three times. The number of private institutions grew at a high speed, starting from 0 in 1989 and reaching 67 in academic 2003/2004, while the number of public higher education institutions increased from 44 in academic 1989/1990 to 55 in academic 2003/2004 (table 9.1).
2. Increase in student enrollments: Total student enrollments grew 3.6 times between 1990 and 2004. As of 2003, around 23 percent of all students in Romania study in a private higher education institution. This percentage has been higher previously: around 40 percent between 1993–1995 and around 30 percent between 1997 and 1999 (table 9.2).
3. Shift in the structural offer of higher education services according to changes in the structural demand. In the early years of their existence, private institutions responded better than public institutions to the structural market demand. While prior to 1990 engineering fields were in high demand, after 1990 engineers were in over-supply. Demand shifted toward qualifications found in shortage at the beginning of 1990 such as economists (needed to assist in the shift toward a market-oriented economy) or jurists, and lawyers, or away from qualifications that at a given point abounded on the labor market. In response to fluctuating market demand, the private sector continually adjusted the offered number of study places in given fields. For example, law enrollments in the private sector decreased from 39.2 percent in 1996 to 23.4 percent in 2003 due to the existence of a large number of law graduates with no jobs while economic enrollments in the private sector increased from 35.2 percent in 1996 to 49.3 percent in 2003 in response to continuing high demand.

This tendency of private higher education to offer fields in high demand such as business, management, marketing, and IT is also common in other countries of Central and Eastern Europe (Kolasinski et al. 2003; Pirozek

Table 9.1 Public and private institutions of higher education, and their faculties, in Romania, 1989/1990–2003/2004

University Year	*Total Higher Education*		*Public HEIs*		*Private HEIs*	
	Institutions	*Faculties*	*Institutions*	*Faculties*	*Institutions*	*Faculties*
1989/1990	44	101	44	101	0	0
1990/1991	48	186	48	186	—	—
1991/1992	56	257	56	257	—	—
1992/1993	62	261	62	261	—	—
1993/1994	63	262	63	262	—	—
1994/1995	63	262	63	262	—	—
1995/1996	95	437	59	318	36	119
1996/1997	102	485	58	324	44	161
1997/1998	106	516	57	342	49	174
1998/1999	111	556	57	361	54	195
1999/2000	121	632	58	411	63	221
2000/2001	126	696	59	438	67	258
2001/2002	126	729	57	465	69	264
2002/2003	125	742	55	489	70	253
2003/2004	122	754	55	513	67	241

Source: Învăţământul în România—Date Statistice, Institutul de Statistică, Bucureşti, 2004, pp. 3, 14–18 (Education in Romania—Statistical Data, National Institute of Statistics, Bucharest, 2004, pp. 3, 14–18).

Table 9.2 Number of students enrolled and existing teaching staff in Romanian higher education institutions, 1989/1990–2003/2004

University Year	*Total Higher Education*		*Public HEIs*		*Private HEIs*			
	Enrollment	*Teaching staff*	*Enrollment*	*Teaching staff*	*Enrollment*	*Percent*	*Teaching staff*	*Percent*
1989/1990	164507	11696	164507	11696	0	—	0	—
1990/1991	192810	13927	192810	—	—	—	—	—
1991/1992	215226	17315	215226	—	—	—	—	—
1992/1993	235669	18123	235669	—	85000	36	—	—
1993/1994	250087	19130	250087	—	110880	44	—	—
1994/1995	255162	20452	255162	20452	114500	44.8	—	—
1995/1996	336141	22511	250836	19994	85305	25.3	2617	11.6
1996/1997	354488	23477	261054	19897	93434	26.3	3580	15.2
1997/1998	360590	24427	249875	21633	110715	30.7	2794	11.4
1998/1999	407720	26013	277666	22955	130054	31.8	3058	11.7
1999/2000	452621	26977	322129	23809	130492	28.8	3158	11.7
2000/2001	533152	27959	382478	24686	150674	28.2	3273	11.7
2001/2002	582221	28674	435406	25174	146815	25.2	3500	12.2
2002/2003	596297	29619	457259	26029	139038	23.3	3590	12.1
2003/2004	620785	30137	476881	26400	143904	23.1	3737	12.4

Source: Învăţământul în România—Date Statistice, Institutul de Statistică, Bucureşti, 2004, pp. 3, 14–18 (Education in Romania—Statistical Data, National Institute of Statistics, Bucharest, 2004, pp. 3, 14–18).

and Heskova 2003) and in other regions (Latin America, Castro and Navarro 1999; Bernasconi 2004).

Private higher education in Romania grew much faster than it did in other Central and Eastern European countries at the beginning of the 1990s. Private student enrollments in Bulgaria, Hungary, Slovakia, Russia amounted to less than 15 percent, much lower than the Romanian 30–40 percent (Malova and Lastic 2000; Davydova 2000; Suspitsin 2003; Caplanova 2002). Toward the end of the 1990s, comparable situations to that in Romania were seen in Estonia, Poland, Ukraine where the size of private higher education was also large, accounting for more than 20 percent of total student enrollments (table I.1 in Introduction Appendix).

A number of factors stimulated the fast development of private higher education in Romania (Nicolescu 2001a):

- the existence of growing demand for higher education at the beginning of the 1990s combined with the inability of public universities to quickly respond to it due to their financial and organizational incapacity (Edinvest 2000);
- the lack of legislation regulating the establishment and functioning of private universities; private entrepreneurs perceived this factor more as an opportunity than a constraint based on the principle "if it is not forbidden, it is allowed;"
- the study fields that were in high demand after 1990 (economics, business, law, journalism, etc.) did not require complex material base, special laboratory equipment and consequently expensive accoutrements. Private initiative appeared mostly in these fields as was the case in other Central and Eastern European countries as well as in Latin America and other regions (Altbach 1999; Mabizela 2004; Levy 1986).

In the context of a legislative vacuum for private higher education, the first private universities had the status of profit-oriented limited companies. Only later on, others were set up as nonprofit foundations that envisaged to offer an alternative to the existing public higher education institutions (Brătianu 2002). The openly stated for-profit purpose of the first wave of private universities (as opposed to the stated educational mission of the public universities) was consequently perceived to be in service of the institutional owners rather than the students. As a result, serious concerns regarding the quality of teaching and learning in these institutions were raised. In addition, private higher education institutions had underdeveloped material base of their own and learning conditions were initially precarious. Moreover, since these institutions depended on a high number of students, they were perceived as easily accessible in comparison with the public universities where tough entrance examinations had to be passed. Consequently, private institutions got the renown of "easy to get in." Finally, these institutions lacked their own professional body and for that reason they hired (in the best case) academicians from public universities who would work part time in private universities, similar to Poland (Kwiek 2003) and Bulgaria (Slantcheva 2002), or they hired underqualified but well connected teaching staff, similar to some African countries (Mabizela 2004; Otieno 2004).

As yet another consequence, due to the reliance on public universities' teachers working part-time, a massive transfer of organizational models, curricula, and teaching methods from public to private universities took place. The above developments, together with the limited exposure of Romanian academicians to foreign educational models and the relatively low degree of entrepreneurship of the educational sector in general, led to the private institutions' gradual transformation into duplicates of the public ones, leaving little room for innovation, initiative, and higher quality education (Dima 1998; Nicolescu 2002). This situation might lead one to believe that private higher education in Romania has not been a serious alternative to public higher education (Mihăilescu 1996; Miroiu 1998; Chiriţoiu and Horobeţ 1999).

However, within the private sector, a great degree of heterogeneity has developed (Nicolescu 2002) in terms of educational practices. At the same time, most of the private institutions preserved their for-profit orientation despite the new Romanian legislation requiring that all higher education institutions, either public or private, be nonprofit. The "very profit-oriented" image of private higher education, inherited from the sector's early development, negatively affected its legitimacy. The lack of confidence in the sector persisted for many years leading to a negative legacy even when conditions strongly improved. In contrast, the public higher education system, the benchmark for the private sector, inherited its traditional positive image of a "quality higher education" as the sector that ensured the communist economy with highly-qualified work force. As a consequence, private providers still have to convince stakeholders that they can offer quality education while pursuing economic goals. Their legitimacy is still questioned.

Social Legitimacy

Literature identifies role typologies for private higher education in different regions of the world (Levy 1986, 1992, 2002b and Geiger 1986, for Latin

America) including Central and Eastern Europe (Tomusk 2003). In Romania, private higher education is expected to have two sets of roles:

- general roles common to the entire higher education system. These would be the roles typically expected from public universities, the ones to which private institutions are constantly compared to. Such roles include offering educational services and market-required skills and knowledge to their graduates.
- specific roles that apply solely to private institutions of higher education. Such roles include increasing access to higher education and fostering competition on the higher education market.

In this context, and following Geiger's (1986) typology of private higher education roles, Romanian private higher education plays, above all, the role of a provider of "more" higher education by trying to respond to increased demand. There are few private institutions that would be qualified as distinctive from the existing public institutions but no "better" ones. The sector had a recognized major demand absorption function for the first 10 years of its existence. At present, with the increase of access at public institutions through privatization (introduction of more student seats based on tuition fees), the role of private higher education is increasingly including the accommodation of less capable students in addition to their excess demand absorbing function.

In sum, Romanian private universities in need of recognition aim at responding to most public expectations of what higher education is meant to be (Reisz 2005). They attempt to acquire social legitimacy by fulfilling both roles generally expected from public higher education and those specifically expected from private institutions. As a result, the social legitimacy of private higher education seen through the perspective of their roles has not been strong. While private institutions have responded well to the specific expectations of their role (such as increased access, competition), they have not been successful in fulfilling the more general roles, expected also from the public sector (such as offering market-required skills, high quality educational services).

Legal Legitimacy

Legal legitimacy is seen here as a two-fold concept: (1) as recognition by the law and (2) as compliance with legal provisions in the higher education sector.

The first law on higher education concerning the private sector was passed in 1993, three years after the establishment of the first private institutions. This was the accreditation law that ensured legal recognition to the institutions complying with the newly-created legal framework.

Compliance with legal provisions refers to external quality assessment measures enforced centrally. In Romania, external quality monitoring takes place only through accreditation, as one form of external quality monitoring

(Harvey and Askling 2003), that pertains to both public and private higher education. Accreditation is seen as a means of legislative recognition that can contribute to legitimacy building. For the private sector, the main role of accreditation is to establish legitimacy through acquiring the quality label. In Romania, as in most countries of Central and Eastern Europe, accreditation is conducted mostly by national agencies that license higher education institutions and validate their functioning by checking their compliance with given minimum standards (Nicolescu 2001a; Slantcheva 2002).

Law no. 88/1993 on the Accreditation of Higher Education Institutions and the Recognition of Diplomas appeared rather as a response to the market evolution than as part of higher education policy. This can lead one to the conclusion that private higher education in Romania initially obtained some degree of market legitimacy, which was only then followed by some degree of legislative legitimacy, that was triggered by the former. The accreditation law aimed to regulate the growing private higher education sector that was highly heterogenous, creating confusion in the higher education system, on the one hand, and opportunities to get an easy diploma, on the other. However, although the law was passed in 1993, it was not enforced before 1996. According to this law, all higher education institutions functioning on December 22, 1989 (only public universities) received accreditation by default. All other institutions established after that date were subject to an accreditation process,[1] conducted by the National Council for Academic Evaluation and Accreditation (NCAEA) through specialized commissions. Such a discriminatory treatment of the two higher education sectors in favor of the public sector was based on the fact that accreditation criteria followed the public universities' model seen at that time as the standard in higher education. It was not before 1999 that this law was amended (through Law no. 144/1999)[2] to introduce periodical evaluation (every five years), seen as a reaccreditation process, for both private and public higher education institutions. As of 2005, a special law on quality of education is being prepared to be promulgated in Romania. The law sets higher standards for the quality of teaching and learning for both public and private institutions of higher education. The new standards follow closer standards from the Western universities.

Although the accreditation process started as a means to control quality and to grant legality and legitimacy to private higher education in Romania, its implementation was a challenge for the authorities. The arbitrariness in criteria selection and the subjective character in setting the minimum standards lead to a conformity exercise in the quality monitoring process. According to Brătianu (2002), these reasons could explain the irrational extension of the private sector even after accreditation was introduced. The imposed self evaluation process, as part of accreditation, degraded to an exercise in being seen as doing something rather than actually doing it, resulting in an increased fabrication management with little or no influence on quality (Blackmore 2002). Moreover, similar to other countries in the region (Russia: Suspitsin 2003), the existence of formal and informal relational networks between top managers of some private universities and persons with political influential positions and academicians participating in the evaluating commissions, contributed to the speed up of the accreditation for selected institutions. Finally, the accreditation system emphasized quantity rather than quality in private higher education focusing mainly on input criteria (such as number of classes, number of academicians, aspects that are of quantitative nature) with almost no reference to output criteria (such as passing rates, employment rates of more qualitative nature) (Korka and Nicolescu 2005).

However, the Romanian accreditation law has been recognized as the first of its type in Central and Eastern Europe. Credit has gone to it for initiating the quality assurance process in the region and for "being the single most influential document ever published in East European higher education" (Tomusk 2004). Accreditation led to improvement in private higher education institutions with respect to quantitative, more easy to measure aspects, such as material base and number of academicians. As of the end of 2005, 27 private universities—out of 67 operating, were fully accredited. From the remaining ones, some universities are under review for accreditation, functioning with only temporary authorization and some others even though functioning with temporary authorization, did not apply for accreditation (as a second phase in the process).

The initial lack of a regulative framework had adversely affected public trust in the private sector. Those private universities that received accreditation later on used it to build credibility and to attract students, promoting it as the official recognition of their compliance with quality standards. Thus, accreditation ensured legal legitimacy and was used to complement the market legitimacy of private higher education institutions. At the same time, the law allowed for the elimination of those private institutions that could not meet minimum requirements. In this way, the law helped in increasing public trust in the sector that was now "guarded by the state" to ensure minimum quality and learning conditions comparable to those in the public sector.

In complying with accreditation standards, geared toward public institutions, private higher education again compares itself to public higher education. To get authorized and accredited, Romanian private institutions made all the efforts to comply with the legally required standards. This compliance of private institutions of higher education with the unique standards set for the entire education sector represent a form of coercive isomorphism, as defined by DiMaggio and Powell (1983).[3] Further on, mimetic isomorphism has been manifested also in Romania as private universities try to copy public universities—developments found in Latin American countries as well (Bernasconi 2004; Castro and Navarro 1999).

Market Legitimacy

Many private universities that have not yet been fully accredited (have not acquired legal legitimacy) legitimate themselves through the market. Higher education services have to be sold twice: first to prospective candidates by attracting them to study in a given institution and second, to the labor market when graduates look for employment. Thus, prospective candidates, students and graduates, and employers are important sources of market legitimation for the private sector. Their acceptance of private higher education contributes to ensuring market legitimacy. Several studies conducted in Romania have shown that private higher education recieves legitimacy from various constituencies but is constantly placed behind the public sector. Private institutions are legitimated by prospective students but are usually ranked second, having a lower degree of legitimacy than public universities considered to be the standard in terms of quality of education. Săpătoru (2000) analyzed the determinants of the choice of institutional form, testing the hypothesis that income and academic ability constitute significant factors in the choice of public versus private higher education in Romania. The commonly held belief that public education is better than private education was sustained by the data in her study where high school pupils who perceived public institutions to be better than their private counterparts far outweighed those who thought private institutions were better, controlling for field of study. According to these findings, private higher education remains the second best choice for most high school graduates based on reputation. In addition, this study revealed that low income—high ability pupils were more likely to choose public education, while high income—low ability pupils were more likely to choose private education. According to Brătianu (2002), the quality of the educational process is determined by both the teachers and the students. In this context, the propensity of low ability pupils to go to private universities may lead to the conclusion that the quality of education at these institutions is not on par with the public institutions.

The lower degree of legitimacy of the private sector, as compared to the public one, has been further documented by recent student data. The introduction of user payers (students who pay for their studies) in public universities created a transformation in the market relationship between public and private institutions in two specific ways (Nicolescu 2005):

1. Many students would rather go as fee paying students to a public university than to a private one. As a result, the privatization of the public universities takes place at the expense of the private ones. During the period 2000–2004, a 37 percent increase in total student enrollments took place, while the proportion of private higher education graduates decreased from 29.5 percent in academic 1999/2000 to 23.1 percent in academic 2003/2004.
2. The increase in the number of fee paying master programs offered by public universities, where access is easy (based on student portfolio, not on entrance exams), induced the so-called phenomenon of "washing of diplomas," when graduates of private universities would attend a master program in a public university in order to get higher credibility and to have a last degree from a renowned public university.

Another study (Săpătoru et al. 2002), encompassing three countries in Central and Eastern Europe, analyzed whether graduates from public universities were more successful in careers than their peers from private institutions based on the probability of finding employment after graduation and the conditions of employment. According to the findings of this study, in Romania, only 64 percent of graduates were actually working after six months after graduation (68 percent from public universities and 60 percent from private universities), with graduates from the public sector having a slightly higher monthly salary (US$ 92) compared to the salary of their private peers (US$ 89) in 2001. From the legitimacy point of view, the fact that graduates of private institutions have lower wages and are more likely to be unemployed compared to those of public universities, coupled with the high unemployment rate among recent graduates, raises questions regarding the quality of private higher education and of the human capital they attract.

In another study, Nicolescu (2003a) focused on employers' perceptions aiming to see if employers differentiated between graduates of public and private higher education institutions. The findings of the study point to existing differences in perceptions of the two sectors. The main strengths of public higher education (rigorous admission process, seriousness in the educational process, good education, positive image) are seen as the main weaknesses of private higher education while the weaknesses of public higher education (in terms of material base, flexibility, and practical experience) are seen as the strengths of the private sector. Public higher education institutions have tradition and stability, but this also translates into rigidity, while private higher education institutions have flexibility that on the dark side is seen as superficiality. Superficiality is understood as easy access to private higher education (therefore attracting less adequate candidates-in terms of their abilities) and as easy getting through the process (classes are very large—100–300 students/course, professors do not always conduct classes, they are too permissive at examinations). According to this study, the companies' policies and practices toward private graduates do not differ from those toward public university graduates (as the law requires equal opportunities) but differences in perceptions about the two categories of graduates exist. Here too, employers formally grant market legitimacy to private institutions, but the comparison with public institutions is permanently done, as the latter is closer to what is perceived to be a good standard.

Professional Legitimacy

Professional legitimacy can be analyzed through the perceptions of the academic community (from the public universities) of the daily behavior of private institutions concerning their internal practices (such as managerial practices, recruitment, and so on).

Private universities follow similar collegial managerial systems as public institutions. In spite of the collegial managerial system, in practice, little decision-making is delegated to faculty of private institutions. Even those faculty in managerial positions are often excluded from decision-making, as the actual decision-making power lies in the hands of the founder or owner of the private higher education institution, who usually exercises an autocratic managerial style. There is little transparency in the way decisions are made, especially with respect to financial decisions. According to a national survey of managers of higher education institutions, 77 percent of the private universities' managers stated that they did not know what the budgets of their universities were (Nicolescu 2003b). The lack of transparency over financial aspects leaves room for the founder or owner of the private higher education institution to misuse funds.

Such features of the managerial practices can have a negative effect on the legitimacy of private universities in the eyes of the academic community over public universities, where, even though not completely free of the risk of power abuse, decision-making is generally more democratic and institutions are accountable to the Ministry of Education and Research for the use of public funds. As noted, private universities are often perceived as having a solely for-profit motive and the existing managerial practices reinforce this suspicion.

Faculty recruitment and hiring practices in private higher education institutions present yet another important aspect of the organizational life:

1. In the early years of their functioning, private universities relied on teaching staff that they hired on a part-time basis from public higher education institutions. This becomes important as the use of academicians from public universities helps increase legitimacy through simple cloning (Levy 1999);
2. Later on, with the introduction of more stringent accreditation standards regarding the percentage of full-time habilitated faculty, private universities hired existing public university professors or associate professors who would be close to retiring, offering them higher wages[4] than in the public institutions. Also, willing to be accredited and to reach the required standards in terms of number of academic personnel with certain positions, many private universities created conditions to form systems wherein their academic staff could be promoted faster. For instance, one person could become a full professor in five years given she/he had few years of experience in the production sector but no academic experience—a situation that would not occur in a public university, equivalent to almost fabricating their CVs so that correspondence with accreditation requirements is in place. At present, many private higher education institutions still use (in the legally allowed proportion) well known professors from public universities in order to increase their legitimacy.
3. Currently, the accredited private universities are no longer forced to comply with certain standards in terms of academic body (as these standards have been already fulfilled), and more regular procedures for recruiting and promoting are used, even though there are still signs that preferential promotional systems are used in some institutions (Nicolescu 2003b).

Some of the early human resource practices, especially recruiting and promotion, again fostered a negative image in the eyes of the academic community. The basis on which such images were created started to change only in the last few years.

Other aspects of the academic life in the private sector such as teaching methods or assessments of students also parallel those in the public universities. Given that many private universities rely on academic staff from the public sector and some of them have been set up by academicians from the public universities, many of the academic practices from public universities have been simply transferred over to private institutions. As a result, private universities have often been seen as copies of the public universities. Teaching methods, for instance, continue to employ mainly classical presentation techniques and less interactive techniques. Student assessment also follows the traditional method that involves evaluation of memorized knowledge and not the acquisition of skills or abilities.

The professional legitimacy of private higher education varies as acceptance from the academic community alternates: (1) the majority of the academicians from public higher education institutions manifest a low acceptance of private higher education practices, while those also working in private higher education institutions manifest a higher degree of acceptance; (2) academicians from private higher education institutions feel their institutions are unfairly treated by being asked to conform to the same requirements as the public sector. Statements in support of this claim are those such as "it is a young sector and it cannot be compared to the traditional public sector."[5]

Institutional Efforts Toward Increasing Legitimacy

The efforts that an institution makes in its daily operations can bolster its perceived social and professional legitimacy. Private institutions in Romania have been active in promoting innovative academic practices, focusing on the students, and using proactive marketing methods.

In their academic practices, private universities employed both innovative strategies and copying strategies from their public counterparts. An example of the innovative academic behavior of private institution was the early introduction of new academic programs not employing the strict traditional specialization pattern of the public sector. Such a flexibility driven by market demands created a positive image in the eyes of the public and prospective students for whom the educational offer was more diverse. In addition, private universities are seen as more student centered—an aspect reflected in part in the organizational culture through which academic and non-academic staff is formally asked to listen to students more and to fulfil their requests. Some private universities also introduced formal class evaluation procedures to receive feedback from students (Romanian-American University from Bucharest in 2000).

Private institutions of higher education are also more market-oriented as far as financing and the use of funds and marketing activities are concerned, but less entrepreneurial with respect to disciplines, curricula, and teaching methods. While private higher education institutions use more proactive marketing methods to attract students (distributing leaflets, visits to high schools, newspaper and radio ads, outdoor displays—even in the geographical proximity of public higher education institutions), public universities are still using at most only static marketing methods such as disseminating information on the web page and displaying it within the facilities of the university. Though more market-oriented, the communication campaigns of private institutions are seen by some researchers (Coman 2003) as having a hidden manipulative character, whose content conveys that the future private university student will benefit from an easily obtained diploma, tuition fee discounts, and a pleasant study location as opposed to studying in a professional and demanding academic environment.

Private universities in Romania could have taken the distinctive role (Geiger 1988) and offered services different from those in the public universities, but most of them did not take the opportunity and instead copied many practices from the successful public universities. These developments represented the most important source of isomorphism in the whole Romanian higher education system (Reisz 2003), the purpose being to minimize risks (Levy 2004). However, this mimetic isomorphism has not been completely successful, as private universities did not manage to reach the academic level of public universities. They only tried to become what Castro and Navarro (1999) said of Latin America "a pale image of something they can never become." They have been only partially successful as private higher education services obtained market acceptance from candidates.

Institutional practices of private universities contributed both positively and negatively to their social as well as academic legitimacy: in the positive sense, through their new programs, more flexible and more market oriented at a time when public higher education "had its hands tied;" in the negative sense, through their admission practices or recruiting and promoting practices. The imitation of the practices of the public universities are seen differently by different audiences: the general public appreciates it since the perception is that similarity to public universities in academic practices brings similarities in quality of education. At the same time, the business and the academic community would have expected more innovative practices from a system that had higher academic and financial autonomy than the public system and at the very beginning of its existence had the freedom to innovate.

Conclusions

The main route used by Romanian private higher education institutions in their search for legitimacy is through replicating or partial replication of significant characteristics of the public institutions, leading to a mimetic isomorphism between public and private institutions. Moreover, legal acceptance through getting accreditation is equally important for gaining legitimacy, the common standards for both private and public also pushing toward coercive isomorphism.

However, looking at the legitimacy of private higher education in Romania through its various components and from the perspective of different audiences and constituencies, a few aspects stand out:

- Social legitimacy: the general public still retains the impression of the profit-oriented and rather low-quality education about private higher education. Social legitimacy is still low, but it is positively influenced by the growth in legal legitimacy (the status of accredited university) and in some forms of economic legitimacy.
- Legal legitimacy: the private higher education sector is getting acceptance from the legislature through the accreditation process, as a formalized way of quality control; private universities are using their accredited status as their main strength to promote themselves. However, the lack of transparency in the accreditation process lets room for suspicion concerning the correctitude in granting it.
- Economic/market legitimacy:
  - Students and graduates grant legitimacy to private higher education as they acquire their services getting something of value to them in return: (1) some opt for it as a second best choice, when they do not get accepted to public universities and (2) some opt to go there as being less requiring in terms of efforts than a public university. Economic legitimacy has a mixed structure, according to the benefits sought by different beneficiaries.
  - Employers formally accept private higher education graduates respecting the nondiscriminatory requirements of the law, but they have different and in many instances opposing opinions about public and private higher education graduates and these can influence their human resource decisions. However, they tend to emphasize more the individual skills, abilities and knowledge than on the institutional form attended, when recruiting employees.
- Professional legitimacy: the academic community exhibits different degrees of acceptance: some consider that private universities are inferior due to authoritarian practices, lower student qualifications, and so on, while others (especially those involved in these institutions) consider that private universities conduct good quality education due to their higher flexibility and comparable academic practices as in the public universities. So professional legitimacy is mixed, but generally is inclined toward lower levels of acceptance.

Romanian private higher education institutions have chosen the handy self-preserving option of copying a successful model (a public university) in order to ensure legitimacy. Further on, the process has been encouraged by authorities who have been setting singular quality standards for higher education at national level, these standards being taken mainly from the traditional public university standards. And finally, the public-private "copying mechanism" was in place through the use of the same academic work force that would behave similarly in both public and private universities, generating similar institutional practices.

However, private universities are generally seen as weaker copies of the public ones (notwithstanding exceptions). There is a lack of strong differentiation in the educational goals and structure of higher education institutions from the public and private sectors, reflecting the fact that they have partially a similar social mission. At the same time, there is still a difference in the perception of the economic goals pursued by the two sectors, with the private sector being seen as a more profit oriented sector. Thus, Romania is an example where public-private higher education is tentatively isomorphic, with the private sector not being able to reach the level of the public sector, but getting enough market recognition in order to further exist and function.

Private higher education has social, legal and overall legitimacy, on the one hand, and it is rather low. On the other hand, from the economic or market perspective (counting heavily for students and employers), the legitimacy is higher. Finally the mixed legitimacy obtained by the private higher education according to different audiences in Romania, is reflected in both state and market legitimacy, that are granted to the private higher education but under which the sector is still vulnerable.

Notes

1. The academic accreditation process in Romania consists of two sequential stages, provisional authorization and full accreditation, each of them based on the same criteria but with different required levels of compliance: number of full time teaching staff, number of teaching staff with senior positions, textbooks written by faculty members, investments in material facilities, pass rates of graduating students.
2. The Law concerning the modification and completion of the Law No. 88/1993 concerning the Accreditation of Higher Education Institutions and the Recognition of Diplomas.
3. DiMaggio and Powell (1983) present three categories of isomorphism: a) *coercive isomorphism* given by a common legal environment or other imposition from the state; b) *mimetic isomorphism* expressed in emulating successful organizations in order to minimize risks; and c) *normative isomorphism* that comes from professionalism and professionals who set dominant norms.
4. Besides the immediate financial benefit, a higher wage in the last five years of activity was reflected (according to the calculation methods in mid 1990's) in higher retirement pensions.
5. Interview with the Rector of a private higher education institution in Bucharest, Romania, November 24, 2005.

Bibliography

Altbach, Ph. G. 1999. The Logic of Mass Higher Education. *Tertiary Education and Management*, Vol. 5, No. 2, pp. 107–124.

Bernasconi, A. 2004. External Affiliations and Diversity: Chile's Private Universitiers in International Perspective, PROPHE Working Paper No. 4, http://www.albany.edu/~prophe/

Blackmore, D. 2002. Globalisation and the Restructuting of Higher Education for New Knowledge Economies: New Dangers or Old Habits Troubling Gender Equity Work in Universities? *Higher Education Quarterly*, Vol. 56, No. 4, pp. 419–441.

Brătianu, C. 2002. *Paradigmele managementului universitar* (The paradigms of the universitary management). Bucharest, Romania: Editura Economică.

Caplanova, A. 2002. Public and Private Initiatives in Higher Education: The Case of Slovakia, *ACE-PHARE AC P98—1020-R* Report, Project *Should Free Entry of Universities Be Liberalized? Estimating the Value of Public and Private Higher Education in Central and Eastern Europe.*

Castro, C.M. and Navarro, J.C. 1999. Will the Invisisble Hand Fix Latin American Private Higher Education? In Altbach, Ph. (ed.), *Private Prometheus: Private Higher Education and Development in the 21st Century*. Chestnut Hill, MA: Center for International Higher Education, School of Education, Boston College, pp. 51–72.

Chiriţoiu, B. and Horobeţ, A. 1999. Euro-shape and Local Content: The Bottom Line on Romanian Higher Education Reform. *Civic Education Project, Discussion Series*, Budapest, Vol. 1, No. 1.

Coman, M. 2003. Public Relations: An Instrument for the Transformation and Development of Higher Education. *Higher Education in Europe*, Vol. 28, No. 4, December. Routledge: CEPES UNESCO, pp. 409–419.

Davydova, I. 2000. *University in Transition: University responses to current education reforms in Russia*. Budapest, Hungary: Open Society Institute.

Dima, A.M. 1998. Romanian Private Higher Education Viewed from a Neo-Institutionalist Perspective. *Higher Education in Europe*. Vol. 23, No. 3. Routledge: CEPES UNESCO.

DiMaggio, P. and Powell, W. 1983. The Iron Cage Revisisted: Institutional Isomorphism and Collective Rationality in Organizational Fields. *American Sociological Review*, Vol. 48, pp. 147–160.

Edinvest, 2000. Investment Opportunities in Private Education in Romania. *Investment Opportunities in Private Education in Developing Countries*. Vol. 3: Country Profiles, Report to the Finance Corporation, presented by University of Manchester, Nord Anglia Plc and the Institute of EconomicAffairs.

Geiger, R.L. 1986. Finance and Function: Voluntary Support and Diversity in American Private Higher Education. In Levy D.C. (ed.), *Private Education. Studies in Choice and Public Policy. Yale Studies in Non-Profit Organisations*. Oxford: Oxford University Press, pp. 214–236.

——— 1988. Public and private sectors in higher education: A comparison of international patterns. *Higher Education*, No. 17, pp. 699–711.

Harvey, L. and Askling, B. 2003. Quality Higher Education. In Begg, R. (ed.), *The Dialog between Higher Education Research and Practice*. Netherlands: Kluwer Academic Publishers, pp. 69–83.

Kolasinski, M., Kulig, A. and Lisiecki, P. 2003. The Stratgeic Role of Public Relations in Creating the Competitive Advantages of Private Higher Education in Poland: the Example of the School of Banking in Poznan. *Higher Education in Europe*, Vol. 28, No. 4, pp. 433–447.

Korka, M. and Nicolescu, L. 2005. Final report to the project *Private Higher Education in Europe and Quality Assurance and Accreditation from the Perspective of the Bologna Process Objectives: The Case of Romania*. UNESCO CEPES.

Kovac, V., Ledic, J. and Rafajac, B. 2003. Academic Participation in University Governance: Internal Responses to External Quality Demands. *Tertiary Education and Management*, Vol. 9, No. 3, pp. 215–232.

Kwiek, M. (ed.). 2003. *The University, Globalization, Central Europe*. Frankfurt am Main Peter Lang, Germany.

Levy, D.C. 1986. *Higher Education and the State in Latin America. Private Challenges to Public Domination*. Chicago: University of Chicago Press.

——— 1992. Private Institutions of Higher Education. In Burton Clark and Guy Neave (eds.), *The Encyclopedia of Higher Education*. New York: Pergamon Press, pp. 1183–1195.

——— 1999. When Private Higher Education Does Not Bring Organizational Diversity: Case Material from Argentina China and Hungary. In Altbach, Ph. (ed.), *Private Prometheus: Private Higher Education and Development in the 21st Century*. Chestnut Hill, MA: Center for International Higher Education, School of Education, Boston College, pp. 17–50.

——— 2002a. Private Higher Education's Surprise Roles. *International Higher Education*, No. 27. Spring, pp. 9–10.

——— 2002b. Unanticipated Development: Perspectives on Private Higher Education's Emerging Roles, PROPHE Working Paper No. 1 at http://www.albany.edu/~prophe/, accessed January 2006.

——— 2004. The New Institutionalism: Mismatches with Private Higher Education's Global Growth, PROPHE Working Paper No.3 at http://www.albany.edu/~prophe/, accessed January 2006.

Maasen, P. and Stensaker, B. 2003. Interpretations of Self-Regulations: The Changing State-Higher Education Relationship in Europe. In Begg, R. (ed.), *The Dialog between Higher Education Research and Practice*. Netherlands: Kluwer Academic Publishers, pp. 85–95.

Mabizela, M. 2004. Whiter Private Higher Education in Africa. *International Higher Education*, No. 34, Winter, pp. 20–21.

Malova, D. and Lastic, E. 2000. Higher Education in Slovakia: A Complicated Restoration of Liberal Rules. *East European Constitutional Review*, Vol. 9, No. 3. Summer 2000. USA: NewYork University School of Law and Central European University, pp. 100–104.

Mihăilescu, I. 1996. *The System of Higher Education in Romania*. Bucharest, Romania: Editura Alternative.

Miroiu, A. 1998. (ed.). *Învătământul românesc azi Romanian education today, Iași*. Romania: Editura Polirom.

Mora, J.G. and Vila, L.E. 2003. The Economics of Higher Education. In R. Begg (ed.), *The Dialogue between Higher Education Research and Practice*. Netherlands: Kluwer Academic Publisher, pp. 121–134.

National Institute of Statistics. 2004. Învăţământul în România—Date Statistice, Institutul de Statistică, Bucureşti [Education in Romania—Statistical Data, National Institute of Statistics, Bucharest]. Romania.

Nicolescu, L. 2001a. Contribution of Higher Education in Transition towards the Market Economy: the Case of Romania. In Kari Luihto (ed.), *Ten Years of Economic Transformation.* Vol. 3, Societies and Institutions in Transition, LUT Studies in Industrial Engineering and Management No. 16, 2001, pp. 253–281.

——— 2001b. Evoluţia sistemelor educaţionale în Europa Centrală şi de Est—analiză Comparativă [The evolution of the educational systems in Central and Eastern Europe—comparative analysis]. *Revista de Management Comparat Internaţional* nr. 2/2001 [Review of International Comparative Management] No. 2/2001. Bucharest, Romania, pp. 75–84.

——— 2002. Reforming Higher Education in Romania. *European Journal of Education,* Vol. 37, No. 1, March, pp. 91–100.

——— 2003a. Higher Education in Romania—Evolution and Views from the Business Community. *TEAM: Tertiary Education and Management,* Vol. 9, No.1, March, pp. 77–95.

——— 2003b. The Impact of the Romanian Higher Education Reform on the University's Financial and Academic Management, *Open Society Institute* Final Report. Budapest, Hungary.

——— 2005. Private Facing Public in Romania: Consequences for the Market. *International Higher Education,* No. 39, Spring, pp. 12–13.

Otieno, W. 2004. The privatization of Kenyan Public Universities. *International Higher Education,* No. 36, Summer, pp. 13–14.

Pirozek, P. and Heskova, M. 2003. Approaches to and Instruments of Public Relations: Higher Education in the Czech Republic. *Higher Education in Europe,* Vol. 28, No. 4, pp. 487–494.

PROPHE, January 2006, at http://www.albany.edu/~prophe/

Reisz, R. 2003. Public policy for private higher education in Central and Eastern Europe. Conceptual clarifications, statistical evidence, open questions, HoF Arbeitsberichte 2/2003, Wittenberg: Institut für Hochschulforschung, Martin Luther Universität Halle Wittenberg (Institute for Higher Education Research at the Martin Luther University Halle Wittenberg).

Reisz, R. 2005. Romanian Private Higher Education Institutions: Mission Statements. *International Higher Education.* no. 38 Winter, pp.12–13.

Săpătoru, D. 2000. Public or Private? Post Secondary Education Choices in Romania, *Open Society Institute.* Final Report. Budapest, Hungary.

Săpătoru, D., Caplanova, A. and Slantcheva, S. 2002. Estimating the Value of Public and Private Higher Education in Central and Eastern Europe: Labour Market Outcomes and Institutional Forms in Bulgaria, Romania and Slovakia, Final Report to the *PHARE ACE* P98-1020-R project *Should Free Entry of Universities be Liberalized? Estimating the Value of Public and Private Higher Education in Central and Eastern Europe.*

Slantcheva, S. 2002. The private universities of Bulgaria. *International Higher Education,* No. 28, pp. 11–13.

Statistical Yearbook of Romania. 2000, 2001, 2003. Bucharest, Romania, National Commission of Statistics.

Suspitsin, D. 2003. Russian Private Higher Education: Alliances with State-run Organizations. *International Higher Education*, No. 33, Fall, pp. 15–16.

Tomusk, V. 2003. The War of Institutions, Episode I: the Rise and the Rise of Private Higher Education in Eastern Europe. *Higher Education Policy*, No. 16, pp. 213–238.

——— 2004. *The Open World and Closed Societies—Essays of Higher Education Policies in Transition*. USA: Palgrave Macmillan Publisher.

Chapter Ten
Public Perceptions of Private Universities and Colleges in Bulgaria

Pepka Boyadjieva and Snejana Slantcheva

Introduction

The appearance of private institutions of higher education after the political changes of 1989 signaled a transformation for the entire system of higher education in Bulgaria. Although private funding for education has its history—best exemplified by a large donation for the establishment of Sofia University,[1] Bulgaria's oldest and largest university—the country has no long-standing tradition in private higher education. As in most communist countries of Central and Eastern Europe, no private institution functioned during the almost half a century of communist rule in Bulgaria. And prior to communism, the private higher education sector was marginal and lacked broad acceptance by the academic community, the political elite, and the public at-large. The Free University in Sofia, founded in 1920 and nationalized in 1939, was the only private institution of higher education in Bulgaria prior to the communist take-over (Boyadjieva 2003). In 1938, a special legal order determined the privileged status of state[2] institutions of higher education. It denied private institutions the right to call themselves universities, to offer programs available in state institutions, or to award diplomas in higher education. When seen in this light, the establishment of private institutions of higher education after 1989 represents a radical development in the Bulgarian higher education system.[3]

Private higher education institutions in Bulgaria have always struggled to attain legitimacy. During its short history, the first Bulgarian private university—the Free University in Sofia—had to often deal with negative public perceptions. As one member of parliament stated in 1931, "two words on the so-called Free University, which is a private enterprise. From the

State's perspective, I cannot understand the justification for its existence . . . And when the State grants its graduates diplomas, I ask, on what grounds can a Free University, a private initiative compete with the State university?" (Stenographic Diaries 1931, p. 1433). Six decades later, similar claims can be found circulating in the Bulgarian press. As an illustration, a recent daily newspaper speaks of "overt and covert favoritism for private universities, where children of the people in power study thanks to lower admissions standards, teaching and, above all, evaluation" (Ilchev 2004). The article also notes that these private institutions pose a threat to "the" University, or Sofia University.

Sociologists have employed the concept of legitimacy to refer to social acceptability and credibility of organizations (Scott 2001, p. 59). Legitimacy is not simply a function of law or compliance with legal standards; next to legal legitimacy we also have social legitimacy, which is rooted in public trust. Organizations need public support to survive and prosper. And this is especially true for new or different organizational forms. "Legitimacy itself has no material form. It exists only as a symbolic representation of the collective evaluation of an institution" (Hybels 1995, p. 243). Seen in this light, the image of an organization in the eyes of the public can be taken as an indicator of the organization's degree of social legitimacy.

Academic institutions are no different in their need for social legitimacy. More specifically, as Birnbaum notes, "Universities are based on social legitimation; we are infused with meaning; and we are perpetuated because we adhere to certain norms, values, and ways of thinking" (2002, p. 1). The social legitimation of institutions of higher education is predicated on the societal acceptance of an institution as "academic." In a country such as Bulgaria where the entire history of higher education until 1989 is practically the history of state higher education, people's "expectations related to the 'academic-ness' of an institution" (Reisz 2003, p. 24) have been shaped by the image of the traditional state universities patterned along the Humboldtian research university model, governed by the state and supported by the public purse. As a result, public perceptions of the private universities and colleges are created against traditional perceptions of academic organizations. In this context, the social legitimacy of private institutions, reflected in those public perceptions, confronts the traditional image of academic organizations.

In this chapter, we focus on the public perceptions of the private sector of higher education as a barometer of the social legitimacy of Bulgarian private institutions. Through a representative public survey, we attempt to identify perceptions of private institutions of higher education in the eyes of the public in comparison with state institutions of higher education. The chapter first provides the context of our discussion with a brief overview of the Bulgarian private sector of higher education and follows with a discussion of the public perceptions of the Bulgarian private sector measured through the survey.

The Method

The analysis is based on findings of a representative public survey[4] conducted in 2004 by a Bulgarian polling agency. The survey sample included 1,119 Bulgarians. The survey served as a direct instrument to gauge how private institutions of higher education are seen by the public—a point crucial to legitimacy. Moreover, the survey questions aimed at identifying public perceptions of private institutions of higher education in comparison to state institutions of higher education. The comparative perspective offered the respondents the opportunity to correlate old and familiar institutions and their activities with relatively new ones. Through this approach, we first analyze the symbolic value of the adjectives state and private in higher education. In other words, to what extent is state or private taken as a marker sufficient in itself to determine the individual choice for a school, regardless of specific institutional characteristics such as quality of education, employment of graduates, material base, and others. As a second step, the analysis goes beyond specific positive and negative aspects carried by the state or private label. We study public perceptions of private institutions with respect to a group of indicators, which relate to the quality of education, level of corruption, relevance of the education and training to the needs of the labor market, and degree of student-centeredness. The findings of the representative survey reflect the perceptions of private institutions in the eyes of the general public and serve as a barometer for the social legitimacy that private institutions enjoy.

Development of the Bulgarian Private Sector of Higher Education

In 1991, soon after the regime change and following the Law on the Academic Autonomy of Higher Education Institutions passed in 1990, the Bulgarian parliament recognized the first three private universities of Bulgaria.[5] These institutions operated in a legal vacuum until 1995 when private institutions were officially recognized by the 1995 Law on Higher Education. The Law also determined the rules for the establishment of private institutions and introduced state accreditation for all institutions of higher education. Given the Bulgarian history of private higher education, the incorporation of private provision of higher education in the Bulgarian legislative framework appeared as a "dramatic change," in the words of one official of the Ministry of Education and Science (Janev 2004, p. 72).

While the initial growth of the private sector before the 1995 Law on Higher Education was tremendous—reaching 130 percent in increase of student enrollments between academic 1993/1994 and 1994/1995, it slowed considerably after that (table 10.1). The sector continued to grow until the 1999 Changes and Amendments to the 1995 Law on Higher Education, which demanded increased responsibility and accountability by institutions of higher education, both private and public. Curricular and operational changes undertaken by institutions needed to be justified, new programs and courses had to be externally evaluated and accepted, and enrollment sizes were to correspond to actual institutional capacities. In 1999, one private institution—the Slavic University in Sofia—was officially closed by the Bulgarian parliament.

With this slight interruption, the growth of student enrollments in the private sector has been steady. Increasing enrollment levels in the private sector appear in sharp contrast to the recently decreasing student enrollments in state institutions. The decline in the state sector is a direct result of intended public policy driven, above all, by future demographic predictions. In fact, a recent World Bank policy report found Bulgaria to be "unique in its decline in higher education participation" (after academic 1998/1999) amongst all members (both old and new) of the European Union (2005, p. iv). In academic 2004/2005, the share of private institutions amongst all institutions of higher education was 30.2 percent; student

Table 10.1 Private student enrollments, private institutions of higher education in Bulgaria, and annual growth

Academic year	*Private student enrollments*	*Annual growth (in %)*	*Percentage of all students in Bulgaria (%)*	*Private institutions*	*Annual growth (in %)*	*Percentage of all institutions in Bulgaria (%)*
1992/93	4909		2.51	5		5.75
1993/94	5491	11.85	2.66	5	0	5.75
1994/95	12630	130	5.66	6	20	6.82
1995/96	20272	60.5	8.1	8	33.33	9.09
1996/97	23746	17.13	9.04	6	−25	13.33
1997/98	26177	10.23	10.05	6	0	13.33
1998/99	32494	24.13	12.03	7	16.67	15.22
1999/00	27414	−15	10.49	6	−14.28	13.33
2000/01	27939	1.9	11.31	6	0	13.33
2001/02	28678	2.6	12.56	11	83.33	22
2002/03	30984	8	13.44	14	27.27	27.45
2003/04	32802	5.9	14.36	14	0	27.45
2004/05	39099	19	16.43	16	14.28	30.19

Source: NSI, at http://www.nsi.bg/SocialActivities/Education.htm

enrollments in the private sector amounted to 16.4 percent of all student enrollments. Currently, 16 private colleges and universities in Bulgaria educate 39,099 students (table 10.1).

Bulgarian private institutions of higher education are by law nonprofit organizations registered according to the Law on Higher Education. Institutions of higher education that are registered to offer educational services through a different law are "deemed illegal" in the Bulgarian context as exemplified most recently by the Minister of Education who "turned over 22 local universities to top prosecutor . . . claiming that their activity in the country is illegal. These universities are actually registered as companies under the Commercial Law but claim to be representatives of foreign education institutions and to offer distance education. This, however, does not meet the demands of the Education Law, meaning that all such universities are illegal" (Sofia News Agency, 10 March, 2006, at http://www.novinite.bg/searchnews.php). The private sector comprises diverse institutions including large comprehensive universities offering an array of programs at all degree levels, liberal arts institutions, specialized institutions offering programs from related study fields, and professional colleges in business, finances, economics, agriculture, including one theater college.

The gradual legal incorporation of private institutions of higher education into the regulative framework has accompanied the growth of the sector. With each consequent change and amendment of the 1995 Law on Higher Education, given aspects of the existence of private institutions has received state recognition. Accreditation—both institutional and program—is an important aspect, introduced with the 1995 Law on Higher Education. Accreditation is accorded by the National Evaluation and Accreditation Agency. In the Bulgarian context, accreditation means compliance with state-imposed norms and standards. It is an obligatory process, and failure to acquire accreditation carries punitive measures that can lead to the closing of an institution. Punitive measures for state institutions of higher education include termination of state subsidy and student admissions; for private institutions they may entail termination of student admissions (in Bulgaria, the number of students to be admitted in any institution of higher education, regardless of its state or private status, is approved by the state).

The Public (In)Visibility of the Private Institutions of Higher Education

As an instrument for gauging public perceptions, the 2004 representative survey questioned respondents on the visibility of private institutions, the symbolic value of *state* and *private* and the operational performance of private universities and colleges in comparison to state higher education institutions.

The visibility of the private sector is reflective of its standing in the public eye and signals the possession of social legitimacy.

Survey results revealed that a considerable part of the Bulgarian population is still not able to connect the idea of private higher education with specific private institutions of higher education—68.5 percent of the surveyed cannot name a private university or a college.[6] The four oldest private universities[7] were amongst the "most visible" institutions that were mentioned by the other 31.5 percent of the respondents. The public visibility of private institutions varies considerably based on the respondents' social background. As might be expected, private institutions are better known to younger people, where almost half of the respondents between 18 and 30 years of age can name a private institution; by contrast, only 20 percent of those aged above 60 are able to do that. Second, more people living in the large towns and Sofia (40–50 percent) can refer to a specific institution whereas only 22–24 percent of those living in small towns and villages can. Finally, the level of education and the type of profession also determine the level of awareness of private institutions: 70 percent of the students, 53 percent of the civil servants and the private entrepreneurs and 66 percent of the people with higher education are able to name private institutions where only 14.5 percent of those with secondary education and 30 percent of the workers can refer to a specific private institution.

The Symbolic Value of the State and the Private

There are substantial theoretical and empirical grounds to argue that the transition to democracy taking place in the former socialist countries is basically different from the transition to democracy, which other countries have experienced (Offe 2004). What is more, Bulgaria (together with Romania) is recognized as "a peculiar case" within the socialist block mainly due to its "culture of authoritarian egalitarianism . . . (which) stands in the way of both a market economy and of democracy" (Offe 2004, p. 515). With this in mind, we tried to assess the symbolic value of the state and the private in the concrete sphere of higher education.

The survey respondents were asked to express their level of agreement with the statement: "All institutions of higher education in the country should be state institutions." Five choices were given to the respondents starting from "completely agree" to "completely disagree." Of the respondents, 40.8 percent chose the "completely agree" and the "I more agree than not" options,[8] while those who completely or partially disagreed with this statement totaled 22 percent.[9]

Apparently, the kind of institutional ownership engages public attention also with respect to higher education. The survey results should be placed in the context of the *etatism* predominant in the social inclinations of the Bulgarian population as well as the specificity of higher education. In Bulgaria, *etatism* in the sphere of higher education has long and stable history. It cannot be viewed only as a projection of common *etatist* inclinations,[10] but has specific measures with respect to education. As noted above, the entire history of the Bulgarian higher education until 1989 is practically the history of state higher education. In this context, people have been traditionally accustomed to relying on the state to organize, license, and monitor higher education. However, after the political changes of 1989, the state has lost a considerable share of its primacy in the organization of higher education due to a number of factors. As one factor, there was the reaction against the old way of state-planned education, for the democratic changes "shattered the respectability and credit of 'left-wing' political, cultural and economic ideas as a whole" (Offe 2004, p. 189). The increasing development of Bulgarian civil society and the growing privatization of the economy opened space for private entrepreneurial initiative. In higher education, this space was occupied by private institutions. As another factor, the Bulgarian state, as is the case with other Central and Eastern European states, could no longer afford to support financially all those who wanted to receive higher education and to monitor a rapidly expanding higher education system.

The existence of a total of 22 percent of the population who disagree with the idea that all institutions should be state run (against 40.8 percent who support this idea) marks a significant change in public perception. Whether this change is only temporary or turns out to be a stable tendency will depend to a large extent on the private institutions themselves.

Public opinion on this issue is an exceptionally sensitive indicator of the social-group and status differences in the Bulgarian society. The statistical analysis of the survey results that we conducted points that there is a significant correlation ($p < .0001$) between the basic social status characteristics of the respondents—education, social group, monthly salary, age, type of living place—and the way they view and evaluate private higher education. There are significant differences in support for all-state higher education amongst the different social groups. For example, the larger the town a person inhabits, the lesser support for this idea is exhibited. In addition, the survey results point to a direct link between the level of education of the respondents and their *etatist* inclinations: the higher level of education determines the greater openness to the idea of private institutions. The higher the monthly salary, the more positive people are with respect to private institutions. The age relationship is also apparent, where the younger the people are, the more open they are to private institutions.

Interestingly, the results point to little difference in the opinion of the people working either in the state or in the private sector (40.3 percent of those in the state sector and 39 percent of those in the private sector agree with this statement, while 26.6 percent and 27.7 percent respectively disagree with it). This data supports the hypothesis that *etatism* with respect to higher education has its specificities and cannot be automatically deduced from the more general *etatist* inclinations.

In order to estimate the symbolic value of defining an institution as *state* or *private*, those surveyed were asked to react to an imaginary situation: "Imagine that you are a student applicant. Money is not a problem for you. The program you would like to apply for is offered both by the state and the private institutions of higher education. Which type of institution would you like to study at?" From the three possible responses listed, 55.8 percent declared they would choose a state institution, 10.6 percent—a private institution, but 28.7 percent declared that the type of the institution did not matter and that they would decide based upon other factors. These results show that the label *state* has preserved its high symbolic value but that a realization of the functional, or real, value of private institutions is also on the rise. The data also leads one to believe that Bulgarian private institutions of higher education have a high degree of legitimacy with those 10.6 percent of the respondents who unequivocally choose private institutions. This degree of legitimacy is proportional to the overall private student enrollment percentage.

The fact that close to one-third of the respondents approach the choice of higher education institution from a rational perspective and are willing to compare existing arguments against predominant popular perceptions is very important. These are people who actually give a chance to private institutions. For them, the state or private character of an institution does not in itself "explain" the attractiveness of an institution. From this, one can draw the conclusion that whether a private institution will be a preferred choice amongst the Bulgarian population will depend above all upon the excellence of the institution.

The Image of the Private Institutions of Higher Education When Compared With State Institutions

The creation of a private sector of higher education in Bulgaria represented a break in a social sphere that had been monolithic during its history. In this context, it is understandable that public opinion perceives private institutions in comparison with state institutions. For a private institution to gain legitimacy, it must be perceived as having advantages over existing state institutions. Hence, our analysis focuses on the perceived advantages and disadvantages of private institutions when compared with state institutions.

Those surveyed were asked to express their agreement with the following statements:

- Private institutions respond faster than state institutions to the needs of the labor market through curricular changes that reflect market needs.
- Faculty at private institutions focus more on students.
- Faculty at private institutions are less inclined to take bribes and gifts than those at state institutions.
- The graduates of private institutions find jobs with more difficulties than those at state institutions.
- Studying at private institutions of higher education is easier than at state institutions.

Although these statements gauge perceived institutional performance, none of them is a direct gauge of legitimacy. However, they contain several indicators that emphasize in a comparative perspective the strengths and weaknesses of private institutions of higher education such as quality of teaching and learning, level of corruption, relevance of the education to the needs of the labor market, and student centeredness. Public perceptions on the strengths and weaknesses of private institutions when compared to state institutions serve as a barometer of institutional standing.

As noted above, public perceptions of private institutions, just like legitimacy of which they are reflective, are not monolithic, but differentiated, partial, fragmented, specific. They are a dynamic measure dependent on the social group and status differences in the Bulgarian society.[11] There is a significant correlation ($p < .0001$ and Contingency Coefficient [CC] between 0.2 and 0.35) between main sociodemographic characteristics of the respondents and all indicators for assessment of private higher education.

In the mind of the public, private institutions are more flexible than state institutions. They react faster to the needs of the labor market. The respondents agreeing with this statement are twice as many as those who disagreed with it (respectively 23 percent and 10.1 percent). Residents of Sofia, people with higher education, those practicing a free profession, students, those engaged in the private sector, and those with moderately high earnings were inclined to support this statement.

The perception that faculty at private institutions pay more attention to their students is supported by 24.5 percent of the respondents (15.2 percent disagree with this statement). The most ardent supporters of this statement are citizens of the small towns, young people aged between 18 and 30 years of age, those practicing a free profession, private entrepreneurs, and those earning above US $200 per month. In addition, 22.5 percent of the respondents believe that corruption is less widespread at private institutions than at state institutions (18.3 percent disagreed with this statement). Among those agreeing were Sofia residents, adults between 31 and 40 years of age, people with higher education, civil servants, and those with higher earnings.[12]

The results of the survey point to a prevailing disagreement that the graduates of private institutions find jobs harder than do graduates of state institutions. 20.9 percent of the respondents disagreed with this assertion, while 13.4 percent agreed. Sofia residents and people living in larger towns were more inclined to disagree with this statement than those living in smaller towns and villages. Obviously, opportunities for professional realization are not only more diverse in the larger cities and the capital, but job appointment criteria used by employers are more rational, relying more on the demonstrated individual skills and knowledge than on perceptions and symbolically leaden expectations. In addition, people with higher education and higher earnings, and private businessmen were amongst the most convinced in the competitiveness of the graduates of private institutions.

The findings also illuminate the weak relationship between the perception of successful job placement for the graduates of private institutions and preferences for a private institution. The conviction that graduates of private institutions find jobs harder than do those of state institutions does not seemingly dissuade those willing to study at private institutions: 14.3 percent of those completely agreeing with the above statement and 20.3 percent of those partially agreeing with it state that they prefer a private institution. Most likely, these results are reflective of the nascent state of the labor market in Bulgaria. They also lead to the hypothesis that people judge higher education not only from an instrumental perspective or as a factor in a successful professional career, but also because of its inherent characteristics. Studying at a higher education institution takes up a significant amount of time in the life of an individual. It has value in itself and should be satisfying as such. Hence, it is important that the environment is attractive (where faculty do not take bribes) and that students feel themselves to be the focus of attention (faculty is focused on students).

Finally, the public perception that studying at private institutions is easier than at state institutions is strong. This statement was supported by 31.5 percent of the respondents, double the 16.1 percent who disagreed with it. It is important to note here that this opinion was shared by all social groups. What is also striking is that some of the groups that were in strong support of the private institutions in all other questions have the highest percentage in support of this statement, including Sofia residents, those between 18 and 30 years of age, those with higher education, students, those practicing a free profession, and those with higher earnings. In general, it is difficult to ascertain a direct link between "easy" education and "lower quality" education. Obviously, the respondents also did not make such a connection since most of them disagreed that graduates of the private institutions find jobs harder than do those from state institutions.

Despite this, however, the survey results provoke serious questions. Do people support and choose private institutions because they hope that requirements there are lower and demand less effort toward the completion of a diploma? To what extent does the "easiness" of education at private institutions reflect its quality and competitiveness?

The little percentage differences in the responses to the above statements emphasize one salient finding from the survey data: there is not a discernibly broad gap between the private and state sectors in the mind of the general public. This finding speaks quite favorably for the legitimacy of the private sector.

Willingness to Invest in Education

Finally, as long as private higher education is strongly related not only to admission test results but also to a given tuition fee, it is logical to assume that the choice of private institution will depend on the possibilities and willingness of Bulgarians to invest in higher education. The respondents were asked to address the question: "What annual tuition fee are you prepared to put aside (or loan) if you are certain that your child will receive education that will secure her/him good professional future?" 39.1 percent stated that they would invest no money or a limited amount of money for the education of their child.[13] In total, and despite the low living standards in the country, it seems that a considerable part of the Bulgarians (17.1 percent) declares to put aside annually US $1240 for education, if convinced that the education will be of high quality (if the tuition fee is US $620, new 11 percent of the population are ready to pay it). As expected, the preparedness to pay a given annual fee varies according to the different social groups. Naturally, those with higher income are prepared to put aside more. However, a considerable number of those with the lowest income is also prepared to significantly invest in their children's education. A surprising fact is that Sofia residents are much more hesitant about investing in education, which signals that private institutions of higher education can expect more applicants from the country. The exhibited willingness to invest in higher education, all other things being equal, suggests a strong legitimacy for the private sector, which relies predominantly on student tuition fees.

A Portrait of Those Who Prefer Private Institutions of Higher Education

From the perspective of their perceptions, people who prefer private institutions:

- believe that faculty at private institutions are less inclined to take bribes than their colleagues at state institutions: 43.7 percent of those preferring private institutions agree with this statement against 21.6 percent from among those preferring state institutions ($p < .0001$; CC = 0.385);
- are convinced that faculty at private institutions pay more attention to students: 53.8 percent of those preferring private institutions agree with this statement against 21.7 percent from among those preferring state institutions ($p < .0001$; CC = 0.402);
- believe that private institutions respond faster than state institutions to the needs of the labor market: 44.5 percent of those preferring private institutions agree with this statement against 18.9 percent from among those preferring state institutions ($p < .0001$; CC = 0.297);
- do not share the public perception that the graduates of private institutions find jobs harder than do those who graduate from state institutions: 31.1 percent of those preferring private institutions disagree with this statement against 18.3 percent from among those preferring state institutions ($p < .0001$; CC = 0.270);
- do not support the statement that studying at private institutions of higher education is easier than at state institutions: 29.4 percent of those preferring private institutions disagree with this statement against 12.5 percent from among those preferring state institutions ($p < .0001$; CC = 0.390).

Some Conclusions: Public Perceptions as a Barometer for the Social Legitimacy of Private Institutions

In the rapidly changing modern world, people's demands on schools of higher education increase and become more dynamic. A considerable portion of the Bulgarian population is neither willing to trust old symbols any longer, nor be easily charmed by new ones. As already mentioned, public perceptions, just like legitimacy of which they are a barometer, are not monolithic, but differentiated, partial, fragmented, and specific. On the one hand, the individual's social background, education, professional status, and age can strongly determine one's worldview and openness. On the other hand, public perceptions are not totally subjective, they are determined by the actual characteristics of the educational process, the specificity of the organization of this process, and the academic environment. Public perceptions are thus formed of varying inclinations. What is more, public perceptions of the quality of the private sector may not be indicative of the legitimacy of an individual private institution. Nevertheless, the analysis of the public image of the private sector of higher education compared to state institutions in Bulgaria leads to several conclusions concerning the legitimacy of private institutions. In the first place, the findings point to the existence of considerable room for the strengthening of the legitimacy of private institutions. A significant part of the Bulgarian society gives private institutions a chance. Although still more inclined toward the state sector of higher education, public opinion is not rigid and frozen. It demonstrates an openness to consider the advantages of private institutions. It is very important that the orientation toward a private institution is a positively determined personal choice. According to the survey results, there is a direct link between people's perception of the advantages of private institutions and their preference for them. The "party" supporting private institutions is a "party" of an electoral type—it has its relatively small hard core and a considerably wider and more mobile periphery. This periphery, however, will be activated only if convinced of the positive efforts and, above all, positive results of private institutions. The legitimacy of private institutions is obviously high with the hard core supporters, or those who choose private higher education. This is an important finding bearing in mind that the legitimacy of an organization "need not be conferred by a large segment of society for the organization to prosper" (Pfeffer and Salancik 1978, p. 194).

A significant conclusion, related to the previous one, elicits good news for the legitimacy of private institutions of higher education from the survey data. Not only do private institutions appear to have strong standing with this part of the population that uses them, but the overall responses of those surveyed show only a moderate state-private gap in the mind of the public. Of course, neither of these conclusions denies that private institutions appear to lag behind state institutions in legitimacy. However, they tend to confirm an overall sense that alongside the legitimacy deficits of private institutions, there is also a substantial degree of legitimacy.

Public opinion also points to several strengths and weaknesses of private institutions and can be taken as a strategic guide in the institutional search for legitimacy. For "While legitimacy is ultimately conferred from outside the organization, the organization itself may take a number of steps to associate itself with valued social norms" (Pfeffer and Salancik 1978, p. 196). Thus, it seems that private institutions of higher education have strong support in local areas. Targeting and serving different geographical locations and social groups would pay off in the end. In addition, the survey results revealed the need for private institutions of higher education to make a special effort in becoming publicly visible and recognizable. It is misleading to assume that the public will naturally notice them and will easily recognize their positive attributes. If left alone, this process will be long and contentious.

Finally, the private sector of higher education is apparently perceived as a rival to the state sector. This competition has the potential to stimulate the development of the entire system of higher education. Problems arise when social perceptions associate studying at private institutions as unfair competition because of it being easier. This perception (Ilchev 2004) is widespread amongst the academic community. Apparently, "easier studying" equals "lower criteria for admissions, teaching and evaluation," all of which lead to lower quality of preparation. Thus unfair competition is found at the institutional level (by state institutions who rival private institutions for students) and at the individual level (by the graduates of state institutions who compete with graduates of the private sector for jobs). In a country with strong *etatist* inclinations, it is natural to relate the reasons for unfair competition with weak state control, while its limitation—with the strengthening of the state and centralization. It is still difficult to assume that the existence of unfair competition amongst institutions of higher education might be in direct result of the underdeveloped nature of the labor market where the diploma as a document (regardless of its contents) can be an important factor in professional job placement. Doubts about private institutions as unfair competitors can be mitigated only on the basis of evaluation of the quality of preparation of their graduates.

Indeed, while public perceptions reveal considerable room for the strengthening of the legitimacy of private institutions, it will ultimately depend on the institutions themselves. The door to happiness, according to Soren Kierkegaard—opens from inside out (Kierkegaard 1991, p. 85). It seems that the doors to "institutional happiness" for private institutions of higher education also open from the inside out.

Notes

1. The brothers Evlogi and Hristo Georgievi donated land and a large sum of money toward the building of Sofia University in the early 1880s and late 1890s.
2. The Law on Higher Education identifies two types of institutions of higher education in Bulgaria: state and private.
3. The private higher education sector in Bulgaria, which appeared after 1989, did not expand to large proportions as it did in other post-communist countries (Slantcheva 2005).
4. The survey field work was conducted at the end of October and the beginning of November of 2004 by Alpha Research—a Bulgarian polling agency. It was commissioned and supported by the Center for Research on Higher Education in Central and Eastern Europe, Bulgaria, for the purposes of this and similar

analyses. The survey is representative for the mature population of the country. The survey sample has the following sociodemographic structure: type of inhabited place (Sofia—12.8%, regional center—32.8%; small town—20.2%; village—34.2%); age (18–30—21.5%; 31–40—17.7%; 41–50—19.2%; 51–60—18%; over 61—23.6%); education (higher—12.9%; semi-higher (higher professional offered at colleges)—3.5%; secondary—46.5%; elementary—37.1%); social group (students—3.6%; workers—30.9%; civil servants—11%; those exercising a free profession—1.8%; private entrepreneurs—5.2%; farmers—0.4%, unemployed—16.9%; retired—29.2%); ethnic group (Bulgarian—82.6%; Turkish—10.5%; Roma—6.9%); sector of employment (state—15.2%; private—35.8%, not employed—49%); monthly income per member of household (under 50 leva—15.6%; 50–100 leva—36.1%; 100–150 leva—25.6%; 150–200 leva—9.7%; 200–300 leva—6.9%; above 300 leva—3.9%; for the text, the exchange rate of 1.61 leva per $1 has been used).

5. The New Bulgarian University, the Burgas Free University and the American University in Bulgaria were the first three institutions created with a decision of the Bulgarian Parliament in 1991 (State Newsletter, N 80/1991).
6. In an open question, the surveyed were asked to name Bulgarian institutions of private higher education.
7. The New Bulgarian University, the American University, the Varna Free University and the Burgas Free University.
8. Where 31.3 percent of the respondents choose the "completely agree" option and 9.5 percent chose the "I more agree than not."
9. 8.8 percent completely agreed and 13.2 percent agreed partially. As a special note here, the percentage of those who chose the option "I cannot judge" (22.1 percent) was the smallest of its kind for the whole survey; it was almost twice lower than for the other questions.
10. According to a comparative study of 11 post-socialist countries, Bulgaria is amongst the countries with strongest etatist inclinations (Bernik and Malnar 2004, pp. 61–66; Tilkidjiev, 2004, pp. 84–85).
11. Although not surprising, 38 to 53 percent of the respondents find it difficult to express their opinion on the formulated statements. It seems that the field of higher education is still "unknown land" to some social groups. This fact can present a source of change of the here provided general picture of the public opinion both in a negative or a positive direction.
12. A 2005 sociological study of the perception of corruption at institutions of higher education carried amongst the students of Sofia University—the oldest and largest Bulgarian university—found that the student community of Sofia University finds that corruption at state institutions of higher education is higher than at private institutions (Anticorruption Education . . . 2005, pp. 35).
13. Those willing to invest no money at all were 24.7 percent, while 14.4 declared willingness to invest a limited amount of money.

Bibliography

Anticorruption Education and Civil Counterbalance of Corruption at Sofia University St. Kliment Ohridksi. 2005. Sofia: Foundation Social Dialogue and Coalition 2004 (in Bulgarian language).

Bernik, I. and Malnar, B. 2004. Trust in the State or in the Benefits from it? (Attitudes Toward the Role of the State in 11 Post-Socialist Societies). *Sociological Problems*, Vol. 36, No. 1–2, pp. 55–75 (in Bulgarian).

Birnbaum, R. Fall 2002. The President as Storyteller: Restoring the Narrative of Higher Education, at http://www.findarticles.com/p/articles/mi_qa3839/is_200210/ai_n9094638, last accessed October 2, 2006.

Boyadjieva, P. 2003. The Free University and the Cost of the Right to be Different. *Razum*, Vol. 1(3), pp. 108–128 (in Bulgarian).

Hybels, R. C. 1995. On legitimacy, legitimation, and organizations: A critical review and integrative theoretical model. *Academy of Management Journal* Special Issue (Best Papers Proceedings 1995), pp. 241–245.

Ilchev, I. 2004. Careful with the University. *Sega*. November 26 (in Bulgarian).

Janev, N. 2004. A Comparative Overview of the Legal Framework for State and Private Institutions of Higher Education. In Slantcheva, S. and Pushkina, J. (eds.), *Private Schools of Higher Education: Myths and Realities*. Sofia: Iztok-Zapad Publishers, pp. 71–74 (in Bulgarian).

Kierkegaard, S. 1991. Selected Works in Two Volumes, Vol.1, translated by S. Nachev. Sofia: Narodna Kultura Publishers (in Bulgarian).

National Statistical Institute. 2005, at http://www.nsi.bg/SocialActivities/Education.htm, last accessed October 2, 2006.

Offe, C. 2004. Capitalism by democratic design: Democratic theory facing the triple transition in East Central Europe. *Social Research*, Vol. 71, No. 3, pp. 501–528.

Pfeffer, J. and Salanchik, G. 1978. *The External Control of Organizations: A Resource Dependence Perspective*. New York: Harper and Row.

Reisz, R.D. 2003. Public policy for private higher education in Central and Eastern Europe. Conceptual clarifications, statistical evidence, open questions, HoF Arbeitsberichte 2/2003, Wittenberg: Institut für Hochschulforschung, Martin Luther Universität Halle Wittenberg (Institute for Higher Education Research at the Martin Luther University Halle Wittenberg).

Scott, W.R. 2001. *Institutions and Organizations*. Thousand Oaks, CA: Sage.

Slantcheva, S. 2005. Legitimating the Goal of Educating Global Citizens. *International Higher Education*, No. 38.

Stenographic Diaries of the Bulgarian Parliament, 22 National Assembly, IV regular session, 67 meeting—March 24, 1931, p. 1433 (in Bulgarian).

Tilkidjiev, N. 2004. The New Post-Communist Hierarchies: the "Block" Division and the Status Order. *Sociological Problems*, Vol. 36, No. 1–2, pp. 76–98 (in Bulgarian).

World Bank. 2005. Bulgaria—Education and Skills for the Knowledge Economy: A Policy Note. Executive Summary, at http://siteresources.worldbank.org/INTBULGARIA/Resources/EducationPolicyNote_EN.pdf, last accessed September 29, 2006.

Chapter Eleven

State Power in Legitimating and Regulating Private Higher Education: The Case of Ukraine

Joseph Stetar, Oleksiy Panych, and Andrew Tatusko

Introduction

The explosive growth of private higher education following independence from the Soviet Union in 1991 was one of many assaults against state monopolies and authority. In its epiphenomenal corollary, questions were immediately raised throughout the Ministry of Education and Science as to the *raison d'être* for private higher education; is it a dangerous competitor that will spawn chaos? Or if it is to exist, how must it be regulated (Stetar and Pohribny 1999). However, soon after 1991 the relationship between the state and Ukrainian private higher education became much more complex than the issue of existence or nonexistence. While Ukrainian private higher education made initial gains in legitimacy outside of the state legal and political structures, it has since become subsumed under state-imposed strictures to maintain a status of legitimacy. However, these state-imposed strictures paradoxically delegitimate even as they legitimate private higher education in Ukraine.

For Ukrainian private higher education, the ability to establish its value or legitimacy among the various and diverse communities it serves or hopes to serve requires that its actions and responses are in line with social norms and beliefs (Rennie 2005). Organizations exist within the broader context of national norms and values. Actions with the potential to undermine institutional legitimacy are likely to occur if an organization violates or appears to violate deeply established social norms or values. Within the Ukrainian context, however, the main (de)legitimating role belongs to the state, which traditionally served as the provider, supplier, and legitimating authority in both the Russian Empire and the Soviet Union (with Ukraine being a part of both).

By their characteristic distinctiveness from the state, private higher education institutions do not have the safety net that the state provides public institutions with regard to its legitimacy granting authority (Slantcheva 2004). As Suspitsin argues in the case of Russian private higher education, "the problems of its legitimacy are linked to broader socioeconomic processes of the transition from centralized state control to a free market regime, which typically occur in conditions of conflicting and contested regulatory and normative frameworks" (2004, p. 2). This also applies in case of Ukraine, where the state exerts a heavy hand in legitimating and regulating private higher education through legislation governing finance, licensing, and accreditation processes. The basic paradox in Ukraine is clear: the state needs private higher education for capacity building, yet an historical mistrust of the private sector is fueled by state-imposed restrictions, limited resources, and ideological constraints that continue to feed and perpetuate the tension.

This chapter begins with an examination of the initial growth and the struggle for legitimacy of Ukrainian private higher education through the lens of the regulative and normative framework in Ukraine. Then, it considers how state policy, occasionally benign, often contentious but invariably conflicted, has served to legitimate, regulate, and at times delegitimate private higher education through frameworks of licensing, accreditation, and governance. Next, the discussion focuses on four primary ways in which legitimacy is enhanced in the midst of this tension. The chapter closes with a synopsis of current challenges to Ukrainian private higher education for future legitimacy.

Data Collection

To understand the dynamics at play within the private sector of Ukrainian higher education, between 2001–2004, American and Ukrainian researchers conducted 43 interviews with individuals representing a variety of educational institutions, governing bodies, and public organizations drawn from the five largest Ukrainian cities: Lviv, Odessa, Kharkiv, Donetsk, and Kyiv. In addition, four Ukrainian private universities, different in size, location, and academic priorities, served as case studies thereby providing an in-depth understanding of their interactions with the state with respect to governance, finances, licensing, and accreditation processes.[1]

Initial Growth of Ukrainian Private Higher Education

Private higher education in Ukraine has undergone several stages of development in the last 15 years (Sydorenko 2000). First private institutions emerged during 1991–1992. They rapidly grew in numbers in the next several years. The initiation of accreditation of the newly established private institutions during 1995–1996 led to a certain decrease in their numbers. Private higher education institutions gained state recognition and issued first diplomas during 1996–1999. Thus, Ukrainian private higher education had to gain state-related legitimacy (i.e., the right to exist in the eyes of the state) by undergoing rather complex licensing and accreditation processes.

Ukrainian legislation, similar to that in many other countries, differentiates between licensing and accreditation. Licensing, giving the right to operate as an educational institution, is the first step in the state accreditation process. Licensing of private higher education institutions in Ukraine began in 1993, about two years after the first private institutions actually started operations. At that time, licensing was based on a relatively modest quality assurance process. Currently, both licensing and accreditation are very complex procedures designed to ensure broad institutional and educational quality. As to accreditation, according to Ukrainian law, an institution can only be accredited upon the graduation of its first students. Thus, since private higher education was only introduced in 1991, no institution could be accredited before 1995–1996.

Statistical data demonstrate a progressive increase in the number of licensed institutions during 1990s. At the beginning of the licensing process, in 1993, only 23 institutions were granted state licenses. The number rapidly increased and by March 1996, 71,000 Ukrainian students were enrolled at 123 licensed institutions (Ogarenko 2000). Totally, during 1993–2000, about 200 private higher education institutions were reported to establish legal contacts with the Ministry of Education and Science regarding their possible licensing (Ogarenko 2000, pp. 76–77).[2] By academic 2004/2005 , the number of licensed (and mostly accredited) private institutions amounted to 202 (MESU, 2005).

Licensing and Accreditation: Legitimation and Delegitimation in Paradox

Regulation of the quality of teaching and learning is part of the "regulatory and normative framework" (Suspitsin 2005, p. 2). In Ukraine, as in Russia, accreditation "means the right to confer state-approved and state-recognized academic degrees and indicates that institutions meet state-imposed educational standards for quality" (Suspitsin 2005, p. 11). In its drive for legitimation, licensing and accreditation are imperatives for the private sector: without a license, a private institution cannot legally operate; without accreditation it cannot issue state-recognized degrees and qualifications, while unsuccessful accreditation automatically cancels the license. It is within these two processes that the interactions and tensions between the state and private institutions are perhaps the most intense.

Throughout the world, as new institutions—legitimated by the market—proliferate in the private sector, which is reciprocally legitimated by increased demand for higher education to meet market demand, they may not fall within the framework of traditional state accreditation standards. This supports the observation that initial distrust of private higher education "reduces the quality of human resource inputs, which in turn reduces the quality of the education provided and justifies the distrust" (Galbraith 2003, p. 552). This vicious circularity would not seem to leave much room for private higher education institutions to retain quality faculty, to attract quality students, and to attract investors.

A related issue is that the "accreditation committees draw the majority of their members from the very public institutions to which the private sector poses a competitive threat" (Galbraith 2003, p. 555). This issue perhaps brings the paradox into the sharpest relief. Those institutions that would leverage the capacity and access concerns of the public institutions are the same institutions that are being controlled by the public institutions that ultimately appear to need them, at the very least, in a capacity-building function. But the issue runs deeper into the notion of social and political transition itself.

When the state becomes such a pervasive force in the shaping of education, it is not easy to assume that private higher education will see immediate gains. Rather, existing structures must be used in order for change to occur at all. With deep mistrust and skepticism working as a constraining factor toward a more positive perception of private higher education, the reformation of social and legal milieu is something that takes quite a while to work itself out. Hence, the forces of legitimation and stabilization along with delegitimation and destabilization pair off to form a narrative with numerous twists and turns. In order to clarify the paradoxical tension in more details, we distinguish more clearly between those sources that legitimate and those that delegitimate in the licensing and accreditation framework in Ukraine.

Sources of Legitimation

The first regulations on Ukrainian state accreditation adopted in 1992 were based on a "highly centralized, Byzantine administrative structure" (Stetar 1996, p. 21). They were replaced in 1996 by a less centralized but even more Byzantine system. The main innovation of the accreditation system in 1996 was to shift the primary stage of the accreditation from the Ministry of Education and Science officials to so-called "special councils." These special councils, or specialty/disciplinary groups, were hosted and staffed by the largest state higher education institutions, thus recognized to be the leading institutions in the corresponding field. Thus, the cohort of the largest and most influential Ukrainian state higher education institutions gained serious control over the accreditation process, including initial decision-making and share of financial streams. As a result, each academic field was subordinated to a single state higher education institution (and partially to several other state institutions represented in a special council's board) that actually determined state policy on a national scale. This structure tempted some of the state-owned institutions working in the most "profitable" academic fields (i.e. those of especially high market demand), to claim their exclusive legitimacy within the corresponding fields of training.[3]

The Ministry also retained considerable control over the licensing and accreditation processes through its "Expert Council," organized as an intermediary level for the licensing and accreditation processes within the Ministry. Finally, the interests of the Ministry and the most influential state higher education institutions were counterbalanced in the formally independent State Accreditation Commission (SAC) as the highest accreditation instance making all final decisions about licensing and accreditation (previously the final decisions were adopted by the Ministry board). In this system of administrative checks and balances, special councils represented the interests of the most influential state higher education institutions, while the Expert Council represented the preliminary position of the Ministry. In the case of contradiction, SAC had some latitude for maneuvering and looking for compromise.

The SAC board was headed by the Minister of Education *ex officio* and consisted of ministry clerks, rectors of key state higher education institutions (mostly those who hosted special councils), and a couple of rectors of the most recognized private higher education institutions. This composition of the SAC board and the scope of its functions remain without substantial changes until now. In the summer of 2001, the accreditation procedure in Ukraine was simplified to some extent. The original intention of the Ministry to dismiss all special councils was achieved only in the accreditation procedure but not in licensing. This seems to be the result of a compromise between the Ministry and the most influential state higher education institutions acting as special councils' holders.

Sources of Delegitimation

The complicated and even contradictory procedures described above already show that both licensing and accreditation, initially intended as a ways to legitimation, often worked with the opposite result. Many additional details served to further complicate the process. Thus, SAC was not formally obliged to explain any of its decisions to any higher education institution; moreover, it usually discussed all cases *in absentia* (behind closed doors). Representatives of the higher education institutions under consideration usually waited outside (sometimes they could be invited inside to answer some questions) and had no control over the decision-making process. If an institution disagreed with the decision of SAC, it could only appeal to SAC itself, as there was no independent arbiter provided. In addition, in 1999 SAC adopted an extensive set of strict quantitative requirements, designed in such a way that formally they could be met only by the best resourced state institutions. These complex and difficult regulations led to a destabilizing effect within the accreditation system.

Essentially, the process gave SAC and those state higher education institutions that hosted special councils, a high level of formal and informal influence over the majority of other institutions seeking accreditation. The private sector, in its turn, questioned the very legitimacy of an accreditation process controlled by the state institutions, which had little interest in encouraging the development of the private institutions. While accreditation was and is essential for the legitimation of the private sector, the accreditation process, under the firm control of the state universities, tended to destabilize and thus delegitimate the private institutions. On the surface, this was a clear example of the popular aphorism "putting the fox in charge of the chicken coop."

In this context, the question of how Ukrainian higher education institutions managed to receive accreditation after November 1999 without meeting at least some of the most fundamental quantitative accreditation requirements stands out. One of the possible explanations points to the deep seated corruption in the Ukrainian society. In other words, by setting the accreditation criteria at a level unattainable for a number of higher education institutions (including many state-owned and most of the private), SAC might not have intended to ruin the Ukrainian higher education system or close the majority of the institutions. In accordance with the old cultural tradition, it might have consciously used formal requirements as a bureaucratic cover for informal relations that might be inevitably set between the accrediting organs and the accredited institution to make the process successful. This Ukrainian phenomenon does not necessarily imply giving of bribes or gifts, since in some cases the informal counterinfluence can be substituted with personal relations, connections, and even sympathy between accreditation experts and representatives of the accredited institutions. However, an absolute majority of the respondents of the survey clearly stated (although mostly unofficially) that at least some sort of "informal relation" is necessary to pass Ukrainian accreditation.[4] Obviously, the bribery phenomenon in Ukrainian education can be corrosive and undermine the legitimacy of the private sector, especially for those institutions that seek to establish themselves in the public conscience as corruption free.[5] As Medvedev remarked, "if an American University, having exclusively Nobel-prize-winning teaching staff, would decide to transfer its base into Ukraine, it would not get here even a license (without a bribe, of course), and it would only dream about accreditation" (2000). Hence, short term gains in legitimation means destabilizing the legitimacy of the accreditation process and resulting in further delegitimation of private higher education.

Some positive changes within the accreditation process, however, gradually appeared during 2000–2003. Since the summer of 2000, accreditation fees have finally been made equal for all higher education institutions, and in 2003, SAC significantly lowered some of its most draconian quantitative requirements for both licensing and accreditation. Now for most of the existing private institutions passing licensing and accreditation is rather an unpleasant routine than a struggle to the death—which actually means that some "legitimation compromise" between the state and the private institutions has been achieved in this important dimension.

State Policy: Sources of Legitimation and Delegitimation

While the licensing and accreditation impose a close, albeit tense relationship with the state, other dimensions of state policies may have a different impact. Private higher education institutions in Ukraine generally base their governance policy by trying to develop a healthy distance from central government perceived to be highly corrupt and riddled with communist-era apparatchiks. These efforts provide private higher education with a distinct aspect of not state-related legitimacy as a growing number of the private institutions offer themselves as corruption-free zones.

While such legitimacy management—a process whereby organizations attempt to gain support for their actions (Rennie 2005)—is deemed critical by a number of private higher education institutions, the distinction from the state sector often carries a steep price. As with licensing and accreditation, the regulatory and normative framework is wrought with an inherent tension placing private institutions at odds with the system. Hence, as seen previously, every legitimating mechanism seems to be coupled with an equal and opposite delegitimating mechanism.

Thus, according to the second version of the Law on Education, adopted in 1996, the Ministry of Education and Science now approves the appointments of rectors in all higher education institutions, including those in the private sector. Each rector of research universities is now required to hold the title of Professor and the research degree of Doctor or Candidate.[6] Age requirement is also imposed—a rector cannot be older than 65 years.

To understand the importance of these, we should note that initially rectors of many private institutions were not elected but appointed by the institution's founders. As a result, in many cases rectors of private higher education institutions were not academicians but rather their founders appointing themselves. The new regulation should aim to help guarantee competence and expertise of educational leaders and in effect further legitimate the sector by helping block tendencies toward nepotism and cronyism endemic to many of the private institutions. However, such governmental regulations can also lead to rather delegitimating consequences.[7]

Ukrainian legislation regarding the establishment of educational institutions also legitimates private sector in a rather contradictory fashion. Only state-owned higher education institutions are legally treated as nonprofit. They are established by central or local authorities in accordance with the "Law on Education." Their main statutory goal is to satisfy the educational needs of citizens. Unlike them, private higher education institutions are legally treated as commercial enterprises and are created by private owners in accordance with the Law on Business Undertakings and/or the Law on Joint-Stock Companies. Accordingly, the teaching staff of private higher education institutions is not legally treated as academic personnel and, unlike teaching staff of state institutions, has no legal right to special increased state old-age pension. Therefore, Ukrainian legislation has obviously discriminative policies toward private higher education institutions by relating them to commercial law and thus partially delegitimizing them as educational institutions, called for providing education rather than making profit by all possible means.

On examination, the financing of private higher education institutions in Ukraine can seem oxymoronic. Because financial regulations are seemingly designed to keep the private institutions in the commercial and heavily taxed category, they serve to constrain the private sector's drive for innovation, a key strategy in the drive to broaden constituency support and establish legitimacy. For example, until recently money paid by a corporate investor for someone's education (e.g., employee) should be reported both as the student's personal taxable income and at the same time as the taxable income of the investor (i.e. for both the investor and the student such expenditures were not tax-deductible).

Also, according to the Law, state-owned higher education institutions are not required to have a statutory or capitalization fund, while the statutory or capitalization fund of a private institution must not be lower than the declared annual tuition fee for the scope of students that, according to license, may be admitted in one academic year. This requirement for a statutory fund, while in theory provided a degree of protection for students by ensuring the institution has tangible assets, also effectively restricted opportunities for increase in enrollment in private higher education institutions. Thus, one may find a complicated mixture of legitimation and delegitimation, especially taking into account that in many cases the regulatory framework is applied ambiguously.[8]

The only good news for private institutions is that even more Byzantine-like financial constrains are imposed on the state-owned institutions' expenditures, so that the private institutions may retain a higher degree of financial autonomy, although under much heavier taxation burden.

Especially complicated situation exists with legitimation as educational institutions of different denominational institutions, initially serving the needs of various Ukrainian churches. According to the law, such institutions are not treated as educational institutions at all, instead they are regarded as religious organizations, thus having no right to offer secular degrees and qualifications. The radical solution to this problem is possible only by changing the Ukrainian Constitution that currently declares complete separation of church and secular education, but in the near future it is not realistic to expect such changes. However, some denominational institutions, following old national traditions and western examples, already managed to break this legal barrier and find a way to combine denominational and secular education, thus legitimating themselves both for the state and for the laity. One of them, Ukrainian Catholic University in Lviv (UCU), is commonly considered the best denominational higher education institution in the nation; other denominational institutions, representing both Orthodox and Protestant Churches, are moving in the same direction.

In sum, on one hand, a number of private institutions offer themselves as distinct options to perceived state sector corruption thus aiming to strengthen their legitimacy. On the other hand, the state serves to both stimulate and hamper (and thus both legitimate and delegitimate) the development of private higher education through its various policies.

Alternative Sources of Legitimation

Given the strong regulatory framework imposed upon private higher education, private institutions have actively searched for alternative sources to enhance their legitimacy. Such sources have included social networking, collaboration with the state sector, and affiliation and stratification. They need not and probably cannot be sought in lieu of the state, but rather in addition to the state.

Legitimation through Social Capital

Since legitimacy is based in part on the perception that an organization's actions and responses are in line with broad social norms and values, private higher education institutions can derive significant legitimacy from the students who attend these institutions, their families, and their broader social networks. Data from 2001 collected by a group of Ukrainian sociologists at Donetsk National University reflect general public and professional opinion concerning state and private higher education institutions in Ukraine.[9]

As research findings indicate, Ukrainian academics have highly critical attitude to the current educational situation in the country (Program of Support 2001, pp. 2–4). At the same time, two-thirds of the respondents were confident that the level of education in Ukraine was not lower than that in the Western countries. Educational professionals generally supported the idea of reforms (64 percent). Interestingly enough, leaders of private higher education institutions were more willing to support reforms than their colleagues from state institutions. Lack of finances and the economic crisis were seen as the main obstacle for further development of the higher education system (76 percent). Corruption in the field of higher education was seen as "quite a serious problem" by 50 percent of the respondents, and as "extremely serious problem" by 36 percent. Corruption was reported to take place in the admission process—64 percent of the respondents, distance education—50 percent, educational authorities—50 percent, at the level of educational leaders—42 percent.

These broad concerns regarding Ukrainian higher education suggested a valuable opportunity for the private institutions to strengthen their social legitimacy. Leading private universities in Lviv, Odessa, Kharkiv, Donetsk, and Kyiv have been able to develop a common understanding among their students and faculty as to what constitutes academic corruption and to foster an institutional culture to create relatively clean institutions (Stetar et al. 2005). Data collated in the year 2004 appeared to confirm that the leading private universities were deriving a high degree of market-related legitimacy evidenced by high satisfaction among their students with respect to the specialties offered, employment prospects following graduation and the academic reputation of the private university (Stetar et al. 2005b). Forty-eight percent of the students from Kharkiv private institutions expressed a desire to go abroad for permanent residence, whereas the same figure for the students from Kharkiv state-owned higher education institutions amounted 73 percent (Sydorenko 2000, p. 174). Such findings boded well over the long term for an enhanced legitimacy of private higher education as these students took their place in Ukrainian society. There was evidence suggesting, not unexpectedly, that students attending the private universities from which we drew our data came from a relatively high socioeconomic backgrounds (Stetar et al. 2005b); this also helped to explain their confidence regarding post graduation employment. Although this socioeconomic background suggested that these students may have had greater opportunities to go abroad, their responses seemed to reflect a degree of optimism regarding the future of their country, their own personal economic future, and a satisfaction with the education they were receiving at their private universities. Collectively, their responses seemed to lay the foundation for growing legitimacy of the private sector.

Legitimation through Collaboration with the State Sector

A potential avenue for enhancing the legitimacy of the private sector is through cooperation and greater integration between state and private higher education institutions. Such collaborative arrangements build upon the established legitimacy of the state institutions of higher education and, by the act of association, serve to legitimize the private universities among a broad range of current and potential constituents. Leaders of private higher education institutions usually tend to avoid any clashes with neighboring state institutions. Absolute majority of the Ukrainian private higher education institutions use "niche" strategy, i.e. orient their educational policy toward a limited but comparatively stable and free—or at least not densely occupied—segment of the educational market. The alternative "growth" strategy is unacceptable for most of the private institutions (Sydorenko 2000, p. 180).

There are different forms of cooperation between private and state institutions. At least one state and one private higher education institution organized a joint faculty for distance education, where the private institution contributed with its infrastructure and the state institution with its right to grant Master's degrees (Ogarenko 2000, p. 132). In these instances, links with a state higher education institution confer a measure of legitimacy through association although they raise questions as to what defines a private university (Ogarenko 2000, p. 131).[10]

Another possible way of cooperation between private and state higher education institutions is a creation of a joint legal unit called educational complex. In this case, the private institution formally remains independent, but actually works as a state institution's subdivision coordinating its activity with the host state institution. Yet another example of cooperation between institutions is the relationship between Ukrainian Catholic University (UCU) and Lviv National University (LNU). This cooperation is conducted without any informal or formal subordination. The formal basis of their collaboration is a special cooperative agreement by which UCU and LNU exchange instructors and teaching materials while capitalizing on their respective strengths for mutual benefit and enhanced legitimacy.

For some reasons, the willingness of both state and private higher education institutions to cooperate significantly differs from one Ukrainian region to another. Odessa, the largest city in Southern Ukraine, appears to be one of the most hostile places for private institutions; only one of the seven private institutions initially opened in that city is currently active. Liubov Shirnina, the Vice-Chair of the Department of Education at Odessa Regional State Administration credits this phenomenon to both the especially high market responsiveness of Odessa's state institutions, arising from the traditional entrepreneurial spirit of the city, and the general incompetence of the leaders of local private institutions. Her opinion is that all but one of Odessa's private institutions indeed deserved to be closed

Conversely, Kharkiv seems to be the region where several leading state and private higher education institutions maintain an active cooperation. Moreover, this cooperation is supported and encouraged by the local state administration. The explanation of Liudmila Belova, first Vice-Chair of the Department for Education and Science at Kharkiv Regional State Administration, is twofold. First, she mentions old academic traditions of Kharkiv region that served as "the source of tolerance and wisdom, and also of some discretion, especially at the beginning, when the attitude to private institutions was more ambivalent." Another reason is that, unlike in Odessa, the founders of Kharkiv private higher education institutions were mostly competent, respected, and experienced educators. Kharkiv Institute of Humanities "People's Ukrainian Academy," a local leader in Kharkiv private higher education movement, is much younger than any of its state neighbors, but is already recognized by Kharkiv state higher education institutions as equal among equals; and at least two or three other Kharkiv private institutions have nearly the same high status.

There are also examples of less than peaceful relationships between private and state higher education institutions in Ukraine. Institutions in these more contentious environments express mutual accusations of unfair competitive tricks, including bureaucratic mechanisms and legal pressures, various violations of licensing restrictions, and other legal limitations. The Ministry of Education and Science has not always served as a neutral arbitrator but rather openly provided some state institutions with one-sided advantages. For example, during the admission campaign of 2000, all state higher education institutions received special permission of the Ministry to admit an unlimited number of students willing to pay tuition fees in addition to the limited number of "budget" students and irrespective of the licensed scope of admission. At the same time, private institutions, which always charge tuition rates, were required do so within the limits of their licenses only (Ogarenko 2000, p. 129). However, since the admission campaign of 2001, this permission for state institutions has not been renewed.

Contradictions usually appear when two or more state and private higher education institutions clash over a narrow circle of the most profitable specialties, e.g. law, economics, or management. Mikhail Dubrovsky,

Rector of the private Kharkiv Social and Economical Institute confessed that several years ago when his institution offered law program, he was visited by a local state attorney who openly lobbied the interests of the National Academy of Law. The visitor's statement, according to Mr. Dubrovsky, was simply "we forbid you to train lawyers!" As a result, this private higher education institution was finally forced to stop the law program in order to survive. While such instances are probably uncommon, they reveal the willingness of the state sector to exert its sheer power to destabilize and thus delegitimate the private sector.

Legitimation through Affiliation and Stratification

The earliest organizations of private higher education institutions were formed for mutual support and desire to influence state policy. In 1993, the Rector of Odessa Institute of Management and Law, founded the Ukrainian Association of Private (or nonstate) Higher Education Institutions in Odessa. Now the association, headed by the Rector of the Kyiv-based European University of Finance, Information Technologies, Management and Business (EUFIMB) embraces not only higher, but private education institutions of all levels. An alternative organization, Ukrainian Confederation of Private Higher Education Institutions, was created in the mid-1990s by the Rector of the International Personnel Academy (IPA) also located in Kyiv. Both EUFIMB and IPA are among the wealthiest and the most influential Ukrainian private institutions.

Not all Ukrainian private institutions belong to these organizations. Some institutions maintain membership at one or both of these associations while others prefer to stay close to a neighboring local state institution or conduct independent policy. The main reason for staying independent is that private institutions "do not expect to receive support from those metropolitan structures [the association and the confederation] and instead pay more attention to strengthening links with regional political and economic elites," (Ogarenko 2000, pp. 116–117) which by sheer association imparts a measure of legitimacy.

The respondents expressed their belief in the increase in the stratification in future among Ukrainian private higher education institutions. Some observers remark that currently the wealthiest and most influential private institutions, including EUFIMB and IPA, aim to and already have much more in common with the most influential Ukrainian state institutions rather than with the rest of private institutions. About 20 percent of private higher education institutions with high academic culture have such a stable market position that their future existence is practically secured. Developing private higher education institutions form another group, which, according to the most optimistic forecasts, may join the first, more established group soon. In addition, the third group of private institutions, about 40 percent of the total, is defined as solely making profit by all means and their future while not universally bleak is certainly precarious (Astakhova et al. 2000, pp. 97–98).

Should this projected stratification occur, a bipolar private sector of higher education will emerge. At one pole will be a relatively small group of high profile, high status private institutions occupying comfortable sinecures while at the other pole will be a larger but shrinking number of struggling institutions with declining public legitimacy and increasing government scrutiny.

International contacts and affiliations, especially with western universities, are a valuable means for building institutional support and legitimacy by providing attractive opportunities for students, faculty, and institutional development as well as lending an aura of cosmopolitanism. Private higher education institutions, especially those situated in the largest and most international Ukrainian cities (Kyiv, Donetsk, Odessa, Lviv, and Kharkiv) are especially successful in this regard.[11]

Challenges to Ukrainian Higher Education

While Ukrainian private higher education has made great strides in seeking and securing legitimacy, its place in the nation's higher education system since its emergence in the early 1990s has and will continue to face significant challenges. One such challenge is the increasing demographic gap. Declining birth rates started in the second half of 1980's due to the Chernobyl disaster and economic uncertainties associated with *Perestroika* caused birth rates to plummet. During the 1990's, many Ukrainian kindergartens were closed (some private higher education institutions are hosted in former kindergartens). At present, Ukrainian secondary schools are feeling the effects of the declining birth rate. With this demographic implosion now reaching higher education, demand will significantly decrease while the competition among higher education institutions reaches a peak. Several of the survey's respondents believe only about 20 percent of the currently existing private higher education institutions will survive these demographic and market challenges.

These sharp demographic projections pose particular challenges to the stability and thus the legitimacy of the private higher education sector. Even if the most optimistic case scenarios outlined by the respondents are realized, the private sector will see a sharp rise in market competition both within the private sector as a whole and with the state sector. Private higher education will have to carefully navigate between the Charybdis of intensive competition that threatens the closing or merger of numerous institutions and the Scylla of increased state scrutiny and regulation that is almost certain to follow any significant instability among the privates.

There will indeed need to be a way for the private institutions to navigate the narrow path between the state structures that grant legitimacy and the freedom granted by their distinctive nature apart from those structures. A few alternative methods for legitimation have been mentioned here that are currently operating within the system. But lest we assume that these alternatives are conclusive, the continued discussion of this issue ought to offer an ample foundation from which other untested alternatives can emerge.

Notes

1. *Acknowledgements.* We express our deepest gratitude to all our Ukrainian respondents, representing various educational institutions, governing organs, and public organizations. It is the openness and sincerity of our colleagues that has made this study possible. Their willingness to cooperate proved to be above all our expectations. Our special thanks are addressed to the authorities of four Ukrainian private higher education institutions who agreed to serve as the subjects of case studies: Ukrainian Catholic University (nee Lviv Theological Academy), Odessa Institute of Management and Law, Kyiv Institute of Investing Management , and Kharkiv Institute of Humanities "People's Ukrainian Academy." The complete list of our 43 respondents representing five largest Ukrainian cities from five different regions can be found at http://www.prophecee.net.
2. In 1996, the ministry of Education and Science announced new accreditation and licensing procedures. The same year, the establishment of the new private institutions had been suspended for several months. However, the process resumed by the end of 1996, so that by 1999 the total number of all licensed private higher education institutions was 132, including 5 universities, 7 academies, 86 institutes and 34 other private higher education institutions (Sydorenko 2000, p. 195).
3. As a compelling case in point, in 1996, Ukrainian authorities attempted to impose state monopoly on law and medical schools and struck at the very essence of private higher education legitimacy by characterizing the "negative tendencies" inherent in having private higher education institutions training professionals in "specialties that influence the security of the state and its citizens"—i.e., doctors and lawyers. The goal was to develop proposals ending the training of doctors and lawyers "in the nonstate and non-profile higher education institutions;" their students would have to transfer to the "profile" state institutions. Corresponding private higher education institutions were proposed to become "ad liberum" structural subdivisions of those "profile" state institutions. When this information was disclosed in the *Business* newspaper, it had the "effect of an exploded bomb" (Ogarenko 2000, p. 100). "For the first time an attempt to violate legal rights of private higher education institutions met an organized resistance" (Ogarenko 2000, p. 100) as a result of which the Prime Minister requested the Minister of Justice to analyze the proposal, and finally the entire story was silently smothered. Nevertheless, the battles around private

Ukrainian Law schools continued. In May 2000 the commissions of the special council in the field of Law "quickly visited other higher education institutions that trained lawyers, counted the number of instructors' work-record books and went away, rejecting the institutions' requirements to check the real quality of the teaching process and students' knowledge" (Medvedev 2000). As previously indicated, the quantitative requirements for both the accreditation criteria and the conditions of licensing, were set so that the licenses and/or accreditation of many Ukrainian higher education institutions could be suspended anytime.

4. As an example, one rector of a private institution confessed that his "limit" of bribery (not in the case of accreditation) is a bottle of cognac or $100–200. He admits that this is much less than is "required" for accreditation, but he simply has no resources to give more; if I had, he adds, I would "give." The reason why he generally agrees to pay money—bribe—is, according to his own words: "If I have 'given' $1000, and have solved problems for $5000, why shouldn't I 'give'?"
5. Some Ukrainian education authorities understand the destructive role of bribery and reject involvement in such illegal activity. One vice-rector of a private institution stated that her colleagues from the corresponding special council behaved absolutely correctly with her. There were no hints or talks about extra payments. Thus, her institution apparently never faced the bribery problem in the accreditation process—although it willingly expressed its "gratitude" to experts (in the traditional form of flowers, candies and alcohol) after the process ended successfully.
6. Ukrainian academic system distinguishes the position of professor (similar to U.S. "full professor") and the title of professor, granted by the special department of the Ministry of Education and Science basing on the request of an institution's Academic Council. Private higher education institutions are deprived of the right to grant a title of professor until the institution is fully accredited, and this is very important in the light of the mentioned newly adopted requirement. The holders of this title enjoy better salaries and usually occupy a corresponding position of professor.
7. One example is evidenced in the Ministry's determination that the founding rector of Ukrainian Catholic University with sterling academic and administrative background including a Ph.D. from Harvard was initially deemed to have insufficient academic credentials.
8. Serhiy Dobrovsky, Financial Manager at Kyiv Institute of Investing Management, says that following Ukrainian legislation resembles going under escort: "one step left or right, and you immediately get penalties, fines . . . At the same time, the legislation is not clear, so that it is not easy for financial managers to define the exact duties of the institute concerning its payments to budget. Rules are interpreted in different ways." Taras Finikov, First Vice-Rector at Kyiv Institute of Economics and Law "KROK" (private), adds that there may be various ways to mitigate the taxation burden for private institutions: while total exemption from the profit tax is desirable, "widening the basis of the tax-deductible expenditures and narrowing the taxable basis" would also be helpful. The problem, however, is that even those more moderate measures are, to date, rarely even considered by Ukrainian authorities.

   Katerina Astakhova, Vice-Rector of Kharkiv Institute of Humanities "People's Ukrainian Academy" (private) remarks that her institution has to pay profit tax even for money received from private sources for secondary education (the Institute contains a secondary school in its structure), although according to law all secondary education, irrespective of the institution's form of property, must be budget-financed, and even partially budget-financed institutions must be exempt from the profit tax. "We got it just once, during one month, about five years ago," says she; and then we were told that "there is no money for us in the city budget." On the other hand, Oleksandr Shubin, Rector of Donetsk State University of Economics and Trade, also expressed complaints against state regulations, that, according to him, give private institutions more "financial freedom," particularly by setting various limitations on the amount of teachers' salary at state institutions.

Only one of our respondents, Nina Oushakova, First Vice-Rector of Kyiv National University of Economics and Trade (state-owned), agreed with the existing taxation inequalities between private and state educational institutions. "I want to reply to all private higher education institutions," said Professor Oushakova, "that they must pay those taxes because they got teachers trained at the expense of the state, and got them for free! This way we [i.e. state institutions] indirectly contributed to their development."

9. Data collection was financed by the local Ukrainian branch of the Soros Foundation and the UN Developmental Program. Data are based upon interviews of 562 instructors from 25 Ukrainian higher education institutions located in four different Ukrainian regions and focus-group interviews with various key educational professionals and administrators.
10. There are at least 15 known examples when state higher education institutions became cofounders of private institutions—including the leading Ukrainian state higher education institutions such as Shevchenko National University, Kyiv National Technical University, Kyiv National Pedagogical University, etc.
11. The international contacts of private institutions are diverse and rapidly growing. Some private institutions widely employ international instructors, have numerous exchange agreements with leading European Union and North American universities and are quite successful in raising funds in the US and Canada. Other institutions participate in a growing number of joint regional international educational programs. Numerous private higher education institutions offer internationally recognized diplomas in close cooperation with Western institutions or professional associations.

Bibliography

Astakhova, V. et al. 2000. Private Higher School in the Lens of Time: the Ukrainian Variant. Kharkiv: Kharkiv Institute of Humanities "Peoples Ukrainian Academy". 464 pages, Original in Ukrainian.

Galbraith, K. 2003. Towards Quality Private Higher Education in Central and Eastern Europe. *Higher Education in Europe*, Vol. 28, No. 4, pp. 539–558.

Medvedev, V. 2000. Illegal field of legal education. *Dzerkalo Tyshdnia' "Weekly Mirror,"* 9 (333), March 03. Original in Ukrainian Newspaper.

MESU 2005. Higher Education and Science—the Most Important Spheres of Responsibility of Civil Society and the Ground for Innovative Development. Kyiv, Ministry of Education and Science of Ukraine, Original in Ukrainian.

Ogarenko, V. 2000. *Non-State Higher Education in Ukraine: The First Decade* Zaporizhia: Institute of State and Municipal Governance. 164 pages, Original in Ukrainian.

Program of Support for Creating the Strategy of Educational Reforms. 2001. Bulletin # 2. Kyiv, June–July.

Rennie, K. 2005. An analysis of image repair strategies: A university in crisis. Unpublished Ph.D. dissertation, Seton Hall University.

Slantcheva, S. 2004. Legitimating the Difference: Private Higher Education Institutions in Central and Eastern Europe. Paper delivered at the June 2004 International Conference *In Search of Legitimacy: Issues of Quality and Recognition in Central and Eastern European Private Higher Education.* Sofia, Bulgaria.

Stetar, J. 1996. *Higher Education Innovation and Reform: Ukrainian Private Higher Education 1991–1995.* Kyiv: USIS, p. 21.

Stetar, J. and A. Pohribny. 1999. Towards a New Definition of Quality: Taras Shevchenko Kyiv State National University. In P. Sabloff (ed.), *Higher Education in the Post-Communist World.* New York: Garland Publishing, pp. 163–189.

Stetar, J., O. Panych, and B. Cheng. 2005. Ukrainian Private Universities: Elements of Corruption. *International Higher Education,* Vol. 38, Winter, pp. 13–15.

——— 2005b. Unpublished. *Choice, Expectations and Satisfaction: an analysis of students enrolled in Ukrainian private higher education institutions.* Department of Education Leadership, Management and Policy, Seton Hall University, South Orange, New Jersey.

Suspitsin, D. 2004. Organizational legitimacy of Russian non-state higher education: Concepts, evidence and implications. Paper delivered at the June 2004 International Conference *In Search of Legitimacy: Issues of Quality and Recognition in Central and Eastern European Private Higher Education.* Sofia, Bulgaria.

Sydorenko, S. 2000. *Private Higher Education: Ukrainian Ways in the World Context.* Kharkiv: Osnova. Original in Ukrainian.

Chapter Twelve
Sources of Legitimacy in U.S. For-Profit Higher Education

Kevin Kinser

Introduction

Organizational legitimacy is an important consideration in the growth of private higher education. Unlike most state-sponsored institutions, the social acceptance of the private sector model relies on its ability to justify its very right to exist. Why should there be private higher education? The answer to this question is obviously different depending on the particular context of the country in which private higher education is established, and is significantly related to the perspective of stakeholders toward existing public sector models. As other chapters in this volume make clear, private higher education serves a demand-absorbing function in many countries, but the proportion of private institutions and their share of enrollment vary widely. Similarly, public higher education is the dominant form of higher education worldwide, though the public sector may or may not retain control over the establishment of new institutions or the offering of degrees. Additionally, the role of a public university could be comprehensive or more narrowly conceived, assigning the provision of other education to nonuniversity private providers. The legitimacy of the public sector, in fact, could itself be questioned, with the establishment of private institutions of higher education as a reaction to diminished expectations for and acceptance of a public university monopoly. In any case, the legitimacy of the private sector is often controversial, even as it continues to expand.

The legitimacy challenges faced by the nonprofit private sector are at the heart of this book. For-profit educational institutions face similar challenges, yet must additionally assert their right to operate as a commercial business with profit-generating revenue and an explicit market orientation. It is noteworthy that much of the emerging private sector globally, also has a commercial orientation, even as it maintains—at least ostensibly—legally nonprofit status (Kinser and Levy 2006). Thus, an analysis of the legitimacy of for-profit higher education provides a close look at the challenges faced by a private sector that is largely commercial in practice, if not in name.

This chapter looks at the multidimensional model of legitimacy proposed by Suchman (1995), and applies it to the private, for-profit sector in the United States. The United States has a well-established private higher education sector that is equal, or even in many crucial respects surpasses, in status to public higher education. The for-profit sector, however, shows evidence of rapid growth and is invading previously sacrosanct areas of nonprofit dominance, challenging accepted educational norms with respect to curriculum, mission, faculty roles, and organizational structure (Kinser 2001; Kinser and Levy 2006). This chapter draws on descriptive accounts of the for-profit sector in the United States, and discusses the legitimacy of the sector from the perspective of the higher education community. In doing so, a recent policy debate is highlighted that centers on the role of for-profit higher education in the reauthorization of the important Higher Education Act. Several observations are offered in conclusion that relate legitimacy in the private sector globally to the concepts articulated in this analysis of the sources of legitimacy for for-profit higher education in the United States.

Definitions

Legitimacy often resists strict definition. The definition used here is the one proposed by Suchman (1995):

> Legitimacy is a generalized perception or assumption that the actions of an entity are desirable, proper, or appropriate within some socially constructed system of norms, values, beliefs, and definitions. (p. 574)

From this definition, the legitimacy of for-profit higher education depends on three elements.[1] First, legitimacy depends on what for-profit higher education does or is perceived to do; second, it depends on the socially constructed environment in which for-profits act; and third, it depends on the evaluations by others of the appropriateness of for-profit actions within the socially constructed environment. These elements suggest that there may be more than one possible evaluation of an organization's legitimacy. The actions of the for-profit sector may be understood differently by different groups, and multiple standards of propriety may apply. An organization, moreover, cannot declare its own legitimacy. Even though it is generally spoken of as something an organization possesses, legitimacy must be granted by others.

It is therefore important to identify who the "others" are who assess the for-profit sector and make legitimacy decisions. The role of the state is often emphasized in making determinations of legitimacy, but Levy (2004) points out that much of the private sector has limited connection to the state and often derives its legitimacy from other sources. Here the role of the state as a formal assessor of legitimacy is deemphasized in favor of a normative perspective that is drawn from the traditional higher education community. In this view, the higher education community does not include the for-profit sector (with the partial exception of the regional accrediting commissions, discussed later), rather it represents an idealized outsider community that judges the for-profit sector according to one or more socially constructed belief systems. For example, the higher education community is dominated by a nonprofit ethos, but is conflicted on the role of the market in higher education. In the first part of the chapter, the various dimensions of legitimacy proposed by Suchman (1995) will be discussed from the perspective of the idealized higher education community.

The second part of the chapter discusses the concept of threshold legitimacy by analyzing the policy debate surrounding the reauthorization of the Higher Education Act (HEA)[2] from the perspective of the Career College Association. While the for-profit sector is a recognized component of the education system in the HEA, the 2004 reauthorization debated the extent of the sector's role in the system. The third section of the chapter weighs the legitimacy potential of the for-profit sector in the United States. The final section reintroduces the international dimension and suggests various implications for the legitimacy of the private sector.

Forms of Legitimacy

Suchman (1995) provides the theoretical framework for this chapter. He proposes a multidimensional framework for understanding legitimacy. Three main forms of legitimacy are identified: pragmatic legitimacy, moral legitimacy, and cognitive legitimacy. A discussion of each form and its relationship to the for-profit sector follows.

Pragmatic Legitimacy

Pragmatic legitimacy involves constituent evaluations of the benefits of the organization's policies and practices, as well as a more comprehensive assessment of whether the organization acts in the constituents' best interests. Suchman (1995, p. 578) identifies three types of pragmatic legitimacy: exchange, influence, and dispositional. Exchange legitimacy suggests that the organization provides rather direct benefits to its constituents, operating in ways that create measurable value for them. Influence legitimacy, on the other hand, relies less on formal exchanges of value, and more on establishing a framework whereby constituents believe the organization understands and responds to their concerns. The organization recognizes constituent interests in the organization and grants them the ability to influence organizational actions. Finally, dispositional legitimacy reflects an anthropomorphic view of an organization in which it is seen as sharing constituent values as well as being good, honest, and deserving of trust.

Does the higher education community see the existence of for-profit higher education as being in their best interest? From the perspective of exchange legitimacy, the for-profit sector would have to provide something of value to other colleges and universities. It is difficult to identify such an exchange. Some not-for-profits may see the sector as valuable because the competition has encouraged them to be more efficient or to accommodate adult or part-time learners. In individual instances, the for-profit school may provide transfer students or relieve the not-for-profit of the need to fill a niche in training students for certain vocational fields (Bailey et al. 2003). In general, however, the exchange is more likely to be seen in negative terms. The for-profit sector uses more than its share of government-provided student financial aid, for example, or it has cherry-picked students and programs in easy-to-serve markets. Exchange, then, does not seem to be a productive source of legitimacy for for-profit higher education.

Influence legitimacy may be more relevant to the for-profit sector. Accreditation is significant here because it is a formal process in which for-profit institutions of higher education invite other member colleges and universities to evaluate their policies and practices. Another area important to influence legitimacy is in the participation of academics from the not-for-profit sector as subject matter experts for curriculum design efforts, in actual program delivery as instructional staff and thesis advisors, or as representatives on for-profit governing boards. While these efforts allow "outsiders" the ability to influence the activities of the for-profit organization, they are not particularly open to the general higher education community. Influence by subject matter experts and other representatives of the higher education community may be limited to those who are already dispositionally supportive of the for-profit model. The faculty hiring process at the University of Phoenix, for example, weeds out critics of the institution's academic model and selects those who support the activities of the institution (Kinser 2001). In addition, faculty and staff at for-profits are not represented by education labor unions, such as the American Federation of Teachers, nor do they host local chapters of the American Association of University Professors. Accreditation, however, may serve as a potential source of influence legitimacy, at least for some for-profits. While most for-profits are accredited by national agencies that are dominated by other for-profits, some institutions are accredited by regional accreditors (Kinser 2005), the agencies that serve almost all public and private not-for-profit colleges and universities.[3] Regional accreditation requires for-profit institutions to adopt certain organizational forms (e.g., independent governing boards), policies (e.g., granting standard degrees), and practices (e.g., hiring academically qualified faculty) that are recognized as appropriate by the higher education community. These for-profits that have earned regional accreditation, then, may be able to rely on influence as a source of pragmatic legitimacy.

The third element of pragmatic legitimacy is dispositional. In order to take advantage of this form of legitimacy, for-profit institutions of higher education would have to be seen as organizations of good will, acting in ways that are associated with positive outcomes for students and society at large. Positive outcomes, for example, could be associated with preparing students for the workforce or with fiscal support of the community by paying taxes. As an organization of good will, the for-profit campus might be viewed as playing by the rules, serving nontraditional populations, or maintaining a focus on educating students rather than padding the bottom line. While these evaluations of for-profit higher education have their advocates (e.g., Ruch 2001), dispositional legitimacy for the for-profit sector seems fairly weak. For example, these schools enjoy no reservoir of good will when problems arise; scandals at one institution have caused stock market jitters across the sector (Blumenstyk 2004). In addition, battles over the reauthorization of the HEA serve as reminders that for-profit higher education is not completely trusted. Proposals to change rules in ways that eliminate distinctions between the for-profit sector and the public and private not-for-profit sectors have met with nearly universal resistance from the higher education community (Ward 2004). Dispositional legitimacy, then, is not well-established for for-profit colleges and universities.

Moral Legitimacy

Moral legitimacy reflects the extent to which an organization is perceived as following and supporting societal norms. This form of legitimacy relies on an evaluation that the organization is doing the right thing, and implies its activities are fundamentally altruistic rather than based in organizational self-interest. Suchman (1995, p. 579) identifies four types of moral legitimacy: consequential, procedural, structural, and personal. Consequential legitimacy evaluates organizations in terms of what they produce. Production is not, however, measured by the number of widgets, but by such ambiguous outcomes as quality, value, and appropriateness. Procedural legitimacy focuses on the techniques and strategies used by the organization to accomplish its goals.

Structural legitimacy involves an assessment of whether the organization is well-suited for its activities. Whereas procedural legitimacy is concerned with distinct actions, structural legitimacy is centered on the larger framework that guides overall organizational activity. Finally, personal legitimacy reflects the confidence that is placed in an organization because of the charismatic leadership of a particular individual.

Does the higher education community see the for-profit sector as a fundamentally altruistic organization, following and supporting the norms for an institution of higher education?[4] From the perspective of consequential legitimacy, the for-profit sector would have to be seen as producing strong, high quality graduates who are comparable to their peers educated in public and private not-for-profit colleges and universities. For-profits might argue that their students are well-prepared to get a job after graduation, and that they specialize in a practical education that is valued by employers. Since many for-profits also offer accelerated programs, the value they provide students may also be seen in a positive light. Finally, for-profits may focus their attention on students that the public and private not-for-profit sectors do not typically serve, giving these students options they might not otherwise have. These consequences of a for-profit education, however, are countered by critics who suggest that such narrowly constructed curricula constitute vocational training rather than higher education (Altbach 2001; Kirp 2003). The accelerated schedule may also not provide enough time for an adequate educational experience, and community colleges, for example, arguably serve a similar student population (Bailey et al. 2003). Still, it is not clear that the education provided by the for-profit sector is contrary to what not-for-profits offer, especially at the two-year level. Moreover, the for-profits have a strong argument for the economic benefit to students, assuming their graduates are well-prepared for the job market. The for-profit sector may, therefore, have a claim to consequential legitimacy, despite critics' questions as to the scope and adequacy of their curricula.

Procedural legitimacy picks up on the critics' questions about consequences and asks whether the methods employed by the for-profit sector are appropriate for an institution of higher education. For example, is the part-time faculty model employed by some for-profits an acceptable strategy? Can an accelerated schedule or vocationally oriented curriculum provide students with an experience appropriate for an institution of higher education? The evaluation here is not on outcomes as in consequential legitimacy, but on the way these consequences are achieved. The prominence of the University of Phoenix, for example, suggest for-profits make wide use of controversial methods (Farrell 2003). At least among the regionally accredited for-profits, however, this does not seem to be the case (Kinser 2005). On the other hand, weak faculty authority over the curriculum, high teaching loads, and lack of a liberal arts core do seem to be the norm in for-profit schools (Kelly 2001). The consequential argument that these activities may produce good results is irrelevant in the calculus for procedural legitimacy. The questionable methods employed by the for-profit sector make it difficult to stake a claim for legitimacy on procedural grounds.

The third type of moral legitimacy rests on structural evaluations of the capacity of the for-profit sector to provide higher education. Is a for-profit college or university, in Suchman's words, "the right organization for the job" (1995, p. 581)? Sperling and Tucker (1997) argue vociferously that it is, based on their experience with the University of Phoenix. Ruch (2001) extends their arguments to other corporate providers of higher education, while Ortmann (2001) identifies a compelling set of reasons for why Wall Street is bullish on the capabilities of the for-profit sector to serve a growing higher education market. To gain structural legitimacy, however, the for-profit sector must be able to overcome a substantial barrier that is fundamental to moral legitimacy: altruism, rather than personal (or organizational) gain, must be seen as central to its activities. The structure of for-profit higher education is about making a profit, and as Ortmann (2001) makes clear, profit has been the driving force behind the expansion of the sector over the last decade. Profit and education are not necessarily incongruous. In fact, for-profit business schools were the dominant form of career education in the latter half of the nineteenth century (Miller and Hamilton 1964) and certainly had structural legitimacy at that time. But with the rise of the public high school at the turn of the twentieth century, and the growing involvement of public and private not-for-profit colleges and universities in vocational, business, and career education, the for-profit model waned and the progressive ideal of education as a public trust took its place. From the perspective of structural legitimacy, modern higher education is a nonprofit endeavor, organized for the public good, and supported with substantial public and private funds in order to fulfill an essentially eleemosynary mission. While the recent rise of for-profit higher education may be a signal that these structural assumptions are being challenged (Slaughter and Rhoades 2004), the structural legitimacy of the for-profit sector remains questionable.

The fourth and final type of moral legitimacy is based on a personal assessment of the leadership an individual brings to an organization. The for-profit sector would gain personal legitimacy if a leader of a for-profit college or university had substantial credibility within the higher education community, or if someone with established prestige came forward as a strong advocate for the sector. For example, when Berkeley professor

Jorge Klor de Alva became president of the University of Phoenix, it caused some to take a second look at the sector (Shea 1998). The second look, however, did not seem to result in a reappraisal, and neither the University of Phoenix nor for-profit higher education as a whole benefited from de Alva's leadership. Currently there are no national leaders in the for-profit sector who could serve as a source of personal legitimacy. Suchman (1995) states that personal legitimacy is rare and tends to have only temporary benefits to the organization. Perhaps a charismatic individual could emerge to lend support to for-profit higher education, but such a scenario does not seem promising for long-term establishment of legitimacy in the sector.

Cognitive Legitimacy

Cognitive legitimacy reflects the level of acceptance or awareness of an organization within a given culture or society. It does not involve an assessment or evaluation of the organization's actions or products. Rather cognitive legitimacy notes whether the simple existence of an organization is acceptable, as distinct from whether one agrees or disagrees with its activities. Suchman (1995, p. 582) identifies two types of cognitive legitimacy: comprehensibility and taken-for-granted. Legitimacy based on comprehensibility demands a coherent explanation of an organization that is consistent with the daily experience and expectations of the audience. The organization must be understood for what it is, and recognizable as viable model for its chosen activities. Taken-for-granted legitimacy suggests that the organization is the only available option for the activity, that any other model is unimaginable, and that considering alternatives is absurd. The organization is accepted as the inevitable result of an endorsed activity.

The two types of cognitive legitimacy suggest that the for-profit sector can either be endorsed as one possible model for the provision of higher education (comprehensibility), or be expected to exist as the natural outcome of society's need for post-compulsory education and training (taken-for-granted). From the perspective of comprehensibility, for-profit higher education in the United States exists in a capitalist culture where there are few prohibitions against conducting an activity for personal gain. The for-profit sector has traditionally used this as a rationale for the existence of their schools. In 1964, Miller and Hamilton wrote:

> Why is it honorable and respectable to buy and sell automobiles for a profit—and not to conduct an educational institution for profit? Why is a man honored for building a modern office building and renting the offices therein for profit—and considered suspect for training, at a profit, the office personnel to operate the same offices? Why is it considered admirable by

> some observers to conduct any kind of legitimate business enterprise at a profit, *except that of education*? (p. 81, emphasis in original)

Other versions of this argument label for-profit higher education as taxpaying institutions, while the not-for-profit sector is comprised of tax spending institutions, or that students can best learn to be productive members of the economy in institutions that are themselves part of that economy. Emphasizing that the for-profit model is consistent with the U.S. economic system, and understandable within it, suggest a claim of legitimacy based on comprehensibility. Ironically, even these arguments recognize the not-for-profit assumptions that surround the provision of education, indicating the fragility of the for-profit rationale. The expansion of the for-profit sector, however, is likely becoming an effective counter to these assumptions. Some of the larger schools, for example, have established a nationwide presence, and may have more name recognition than all but the most well-known not-for-profit campuses. The sheer ubiquity of for-profit higher education, therefore, can be considered fairly strong evidence of legitimacy based on comprehensibility.

The second type of cognitive legitimacy—taken-for-granted legitimacy—is much more difficult to achieve. Paradigmatic status is a rare accomplishment for any organization, though, importantly for this discussion, the university model that emerged from medieval Europe enjoys worldwide dominance (Altbach 2003; Kerr 2001). To be taken for granted, the for-profit model would need to either overthrow the broad organizational patterns established by the medieval universities, or to redefine the arena in which it operates such that it becomes the sole model for a new institutional category. The former option seems unlikely. In fact, the for-profit sector often mimics the standard university model through symbolic practices such as commencement, scholastic conventions such as discipline-centered departments, and academic structures such as graduate schools and colleges of education. The latter strategy is occasionally attempted, however, most prominently by Sperling and Tucker (1997). The authors suggest that because the development of the "adult-centered university" required the bottom-line focus of the for-profit University of Phoenix, for-profit higher education should be the exemplar organizational form for adult workforce development. While Sperling and Tucker's educational transformation is not widely accepted, concern that profit-driven decisions are transforming the academy is frequently expressed (e.g., Kirp 2003; Slaughter and Rhoades 2004), though for-profit higher education is seen as a symptom rather than a cause of this transformation. The existence of the for-profit sector may be comprehensible according to this literature, but it is certainly not taken-for-granted.

Meeting the Legitimacy Threshold

Legitimacy is problematic for for-profit colleges and universities. Indeed any nontraditional educational institution will struggle to establish legitimacy because it relies so heavily on status quo evaluations of organizational mission and strategy. What this review of the sources of legitimacy suggests, however, is that for-profit higher education in the United States can claim legitimacy in three fairly distinct ways. First of all, for-profits can claim pragmatic legitimacy through the influence of accreditation, particularly regional accreditation. They can claim moral legitimacy based on the positive consequences of the for-profit curriculum for students seeking employment. And, finally, they can claim cognitive legitimacy because their for-profit model is comprehensible and understood as a possible option for an institution of higher education. Does this mean that for-profits are "legitimate" institutions of higher education in the United States? Not exactly—as noted in the discussion above, each of the claims is problematic. Regional accreditation is held by a minority of for-profit institutions in the United States (Kinser 2005), and so influence legitimacy may be limited to a small proportion of the sector. Consequential legitimacy relies on vocational outcomes that constitute a constrained perspective of the purpose of higher education. Legitimacy based on comprehensibility assumes a general awareness of private, for-profit institutions as organizationally distinct from private, not-for-profit institutions. While these problems do not delegitimize the for-profit sector, they do indicate that conclusions about the legitimacy of the sector are, at the moment, rather tenuous.

The uncertain legitimacy of the for-profit sector may or may not be a problem. Legitimacy is, in essence, evaluated on whether it meets some threshold determination according to the purposes of the organization. Suchman (1995, p. 574) states that legitimacy can aid organizational continuity as well as organizational credibility, and facilitate passive or active support for the organization and its activities. The question, then, is not whether the for-profit sector is legitimate. Rather, the issue is whether for-profit higher education is legitimate enough. Are for-profit colleges and universities legitimate enough to be credibly seen as full-fledged institutions of higher education, or are they simply seeking continuity as legitimate post-secondary training institutes? Is the for-profit sector legitimate enough to be actively supported by the public, or is a lower level of acquiescence to for-profits as a legitimate private sector industry all that is required? To the extent that organizational continuity and acquiescence are the primary drivers, the legitimacy threshold may be easily achieved. On the other hand, a much more substantial legitimacy threshold is often needed for organizations that seek new credibility and active support for their activities.

Simply because the for-profit sector has been growing dramatically does not in itself suggest answers to questions of continuity or credibility, active or passive support, or whether a particular legitimacy threshold has been crossed. There has been, however, a massive lobbying effort by the for-profit sector to influence federal legislation on higher education (Burd 2004) that does provide evidence that is helpful in this analysis. The Career College Association (CCA)[5] has ten legislative priorities for the for-profit sector that relate to the reauthorization of the HEA (CCA, 2004c). It is an illustrative list. Six of the ten items advocate eliminating or revising various rules and policies that distinguish for-profit institutions from not-for-profit institutions in the Act. The most significant legislative priority in this respect is the adoption of a single definition for a "higher education institution" that would be inclusive of the for-profit sector. Other legislative priorities include establishing "non-sector-specific" accountability measures to provide consumer information on institutional effectiveness, eliminating barriers to credit transfer between regionally and nationally accredited institutions, changing a rule that requires for-profit institutions to earn at least ten percent of their revenue from sources other than federal student aid programs (Commonly called the "90–10" rule, because it stipulates that at most 90 percent of revenue can come from federal aid, with a minimum of ten percent coming from student tuition or other private sources), and revising two rules that currently call for additional scrutiny of institutions when they change ownership. Of the four priorities remaining, two involve revisions that, while applicable to all institutions, arguably are disproportionately important for the for-profit sector. One involves changing the rule that limits financial aid available for distance education to institutions that enroll more than fifty percent of their students in classroom-based programs (the so-called 50% rule). The second recommends clarifying the availability of judicial review for Department of Education decisions, which could open a new avenue for challenging negative evaluations of a for-profit school. The final two priorities relate to student aid provisions in the act. One advocates revising the calculation of student aid refunds owed to the government when a student withdraws from school, and the second is a general call for increasing funding for student grants and loans.

Several of these priorities represent fairly direct requests for active support of the for-profit sector by the federal government. The "single definition" rule, in particular, would allow for-profit institutions to participate in many federal programs for which they are currently ineligible. Establishing common credit transfer policies would help for-profit institutions attract new students. And the elimination of the 90–10 rule would permit federal student aid to provide 100 percent of for-profit sector revenue. Passive support for the for-profit sector may be implied by the two legislative priorities regarding change in ownership rules. Both suggest that the buying and selling of for-profit institutions should be accepted as a common business practice rather than be treated as a special circumstance requiring rigorous scrutiny. The general tenor of all the priorities, though, is clearly in the direction of seeking active support for the for-profit sector.

It is less clear from the list whether the sector's priorities indicate the pursuit of continuity or credibility. Continuity implies that the for-profit sector is confident in asserting its traditional vocational training mission. From this perspective, the agenda can be seen as the for-profit sector claiming certain prerequisites in pursuit of that mission. The items that call for additional student aid, revision of change of ownership rules, and judicial review of department decisions reflect this view. Credibility suggests not only that the for-profit sector is worthy of support, but that it can be trusted as well in the same way that not-for-profit private and public colleges and universities are. As a credible organization, then, the legislative priorities represent a claim for equal treatment. Prominent in this perspective would be the single definition rule, elimination of transfer of credit barriers, standardizing accountability provisions, and changing the 90–10 and 50 percent rules. Even though the overall priority list relates both to continuity and credibility, however, a separate "action alert" from the CCA (2004a) identifies the four credibility items as the "Top 4" issues in the reauthorization of the HEA. The arguments marshaled in the alert focus on treating all postsecondary institutions "fairly and equitably" and assuring lawmakers that "fraud and abuse" concerns are unwarranted. Not just the worthiness, but the trustworthiness of the sector is being promoted, suggesting that credibility is the dominant concern of the for-profit institutions.

Legitimacy and Public Policy

The public and private not-for-profit models for colleges and universities are still institutionally dominant and therefore continue to enjoy favored status in terms of government regulation and policy. In the decentralized U.S. educational system, however, the government is a weak arbiter of legitimacy. The state serves more significantly as an amplifier rather than a guide for consensus on these matters. Educational policy out of sync with educational legitimacy will often fail. State Postsecondary Review Entities (SPRE), for example, were established in 1992 to address perceived problems with the oversight of higher education institutions involved in federal aid programs (Lovell 1997). The program was seen as an illegitimate affront to institutional autonomy and independent accreditation, and quickly withered in the face of substantial opposition from the higher education community (Bloland 2001).

Relevant to the discussion of this chapter, a substantial issue in the program's demise was the requirement that all postsecondary institutions—public, not-for-profit, and for-profit—would be equally responsible to the SPRE. The higher education community was not willing to accept an arrangement that treated traditional institutions of higher education the same as "a mom-and-pop dog grooming school with a one-room store-front operation" (Lovell 1997, p. 341). A decade later, another version of this issue is being contested—this time under the "single definition" label. The success that the for-profit sector has had in gaining legislative support for its agenda, may be short-lived unless questions regarding its legitimacy can be resolved.[6]

From a policy perspective, then, for-profit legitimacy might more productively be seen as a threshold question. At the most basic level, for-profits must be legitimate enough to exist. But a different level of legitimacy is needed for the next level, that of acceptance and inclusion in the higher education community. Finally, the threshold to be crossed is at its greatest when for-profit colleges and universities seek enough legitimacy to support an affirmative policy environment. Assessments of the sector's pragmatic, moral, and cognitive legitimacy suggest the first level of existence is clearly supported, and indicate a presumption in favor of the second level of inclusion. There seems to be limited support within the higher education community, contrasted with somewhat stronger support among policymakers, for the third level of affirmative policy. What could tip the balance? Each aspect of legitimacy can be examined to see how likely improvement or decline is.

- *Cognitive legitimacy* is likely to grow stronger. Simply by continuing to offer an alternative model, the for-profit sector becomes increasingly comprehensible as a legitimate option for higher education. Taken-for-granted legitimacy remains difficult, unless the sector can challenge the not-for-profit sector on the definition of profit. Why for example, should the continuing education programs at Columbia University not be considered a for-profit enterprise? Even though many within the higher education community would raise similar questions about the "marketization" of higher education (Kirp 2003), it does not seem likely that this line of reasoning would readily benefit the taken-for-granted legitimacy for-profit sector.
- *Moral legitimacy* shows strengths and weaknesses. Consequential legitimacy is likely to improve, as the job-focused outcomes of a for-profit education match what is increasingly seen as the purpose of higher education. Structural legitimacy is problematic. As the recent trends in the U.S. health care industry suggest, for-profit organizations in traditionally not-for-profit arenas often find their motives questioned even after establishing their consequential legitimacy. Procedural legitimacy is

more unpredictable. Whether the practices of the for-profit sector, especially the more radical methods of the large corporate chains, will become acceptable is not clear. Consumerism or the growth of distance education could drive this issue. As would be evident in many colleges and universities today—experimenting with alternative schedules and on-line learning—it could become harder to say that the traditional models are without peer. Personal legitimacy could improve if a champion of sufficient stature arises—say, for example, the President of the United States places a high priority on the for-profit sector. This is not a particularly likely scenario.

- *Pragmatic legitimacy* also is a mixed bag. From the perspective of policymakers, the lobbying efforts of the for-profit sector suggest that exchange legitimacy has the potential to greatly increase. On the other hand, the higher education community still views the sector as potential competitors for government largess, making a positive exchange for their support unlikely. Influence legitimacy seems prone to increase, especially if the regional accreditation of the for-profit sector continues to expand. Dispositional legitimacy is uncertain. The recent spate of controversies involving the practices of some well know institutional representatives of the for-profit sector (Blumenstyk 2004) remind everyone that this industry has a history of fraudulent activity.

It would not be difficult to look at these brief predictions and see the weight of the legitimacy equation tilted toward increasing legitimacy for the sector rather than declining. This is because a loss of legitimacy is typically the result of some unpredictable crisis, whereas building legitimacy can be a proactive organizational strategy (Suchman 1995). It does not, however, mean that the public policy questions will ultimately be decided in favor of the for-profit sector. As a socially constructed concept, legitimacy is not malleable to organizational ends. The ends, rather, must be molded to accommodate legitimacy. In other words, the for-profit sector could become increasingly legitimate within the system of higher education, but primarily as vocational training institutes. The role of a degree-granting college or university may still be dominated by the not-for-profit private and public sectors. The policy environment, in any case, will echo these socially constructed legitimacy decisions regarding the for-profit sector, serving to reinforce the perceptions of some and challenge the perceptions of others.

International Implications

The development of the private sector in much of the world has substantial parallels with the for-profit sector in the United States (Kinser and Levy 2006).

Legitimacy of these new sectors is dependent on reformulations of the standard university model, including alternative educational practices and organizational forms. Examining the legitimacy of the for-profit sector in the United States following Suchman's (1995) multi-dimensional framework suggests several implications for analyzing the legitimacy of the private sector in international contexts.

1. It is productive to consider the relationship between organizational goals and legitimacy in the private sector. DiMaggio and Powell (1983) place much emphasis on the homogenizing effects of organizational field. The purpose of the new private sector, however, is often rather different from the state-sponsored universities (Levy 2004). In terms of degree level, for example, the bulk of private enrollment tends to be in nonuniversity institutions. In the United States, the parallel for-profit sector is dominated by a narrow, career specific curriculum. The organizational field, in fact, may include much organizational diversity, and a variety of claims on legitimacy, based not on the standard established by existing universities, but by an emerging standard employed by the alternative private sector model. Empirical evidence of this in the U.S. case can be seen in Kraatz and Zajac's (1996) analysis of the curricular diversification of private liberal arts colleges. Increasing vocationalization of the curriculum had no negative effects on the colleges, even as they moved counter to the norms established by the most elite organizations. In other cases, such as those described by Pachuashvili in this volume, the private sector institutions intentionally position themselves in opposition to the dominant public sector. To be different, then, is not prima facie evidence of illegitimacy, and alternative goals may be legitimized within an organizational field.
2. The social construction of legitimacy remains significant, and the private sector cannot ignore the shared assessment of a collective audience. There may be competing assessments, however, as groups offer differing interpretations of the various aspects of legitimacy. To the extent that the assessments coalesce, legitimacy is stronger. To the extent that they diverge, legitimacy becomes questionable. In the United States, various views as to the legitimacy of the for-profit sector have come together in terms of comprehensibility, for example, while they still remain far apart in terms of disposition and structure. In the private sector globally, other aspects of legitimacy may be significant, depending on local conditions.
3. The decline of the state's influence and power, and the general rise in privatization across all sectors, suggest that the state's role is as an

amplifier rather than an arbiter of legitimacy. The U.S. case described here shows how for-profit sector claims of legitimacy are framed in public policy debates, but may not necessarily be decided there. Levy (2004) notes that when the state itself loses legitimacy, organizations build their legitimacy on ties with other actors, including international institutions. This can create a feedback loop where externally derived legitimacy pushes the state to establish policies that legitimate new organizations within existing legal and regulatory frameworks. Private sector institutions can gain legitimacy from their connections with other organizations and then use that legitimacy to achieve formal status with a more reluctant state. In the future, certain transnational for-profit educational providers, such as Laureate or Apollo, may serve in this role. Legitimacy thus derived from formal connections with well-known international organizations may then be amplified by the state by granting recognition or permission to operate. To the extent that gaining legitimacy is considered a strategic organizational activity, such back door methods may become important in international contexts.

4. Influence legitimacy may be a particularly significant dimension in the growth of the private sector. In the U.S. for-profit sector, the example of regional accreditation suggests such legitimacy can influence not only the emerging organization, but also the recognizing entity. In international contexts, there are many examples of public-private partnerships that serve to legitimize the private sector organization (Kinser and Levy 2006), and it is important to recognize that the influence exerted may, in fact, work in both directions. Private higher education often relies on accreditation or recognition from a public entity in order to grant degrees. This relationship has the potential to influence the public sector and encourage the development of more private higher education. It is, in essence, the proverbial camel's nose under the tent. Once one private sector institution is legitimized through an influential relationship, the door may be opened wider for others to become legitimate educational partners, transforming the public sector in the process.
5. The rise of the private sector implies new competition in the higher education marketplace. The marketization of higher education (Kirp 2003) further suggests the potential "sovereignty of the consumer" in determining legitimacy (Gumport 2000, p. 79). From the framework discussed here, this new "sovereignty" privileges pragmatic legitimacy as well as the consequential aspects of moral legitimacy. The pragmatic dimension suggests the consumer will look for an exchange with the private sector institution, seeking something of value for his or her

tuition. Looking to be accommodated, the consumer pursues an educational experience that is convenient, thus influencing the school to revamp schedules and curricula. The personification of the private sector as "better" or more "student-centered" suggests the emergence of a more positive dispositional legitimacy. Finally, the practical consequences of private sector enrollment for the consumer are seen in terms of employment and economic development, diminishing the value of culture and the role of higher education as a social institution. This is, of course, just one potential script. Suchman's (1995) typology, however, implies that some sources of legitimacy may be more or less relevant than others, depending on the audience. As the audience shifts to the consumer, sources of legitimacy are likely to shift as well.

Conclusion

This chapter should be seen as observations on legitimacy in the for-profit sector and the international private sector rather than a formal investigation into the actual sources of sectoral legitimacy. The multidimensional framework allows for a more complex view of legitimacy, and also helps to identify where for-profit higher education has a strong claim to legitimacy, and where it is weaker. For the private sector globally, a similar analysis may be conducted by specifying the actors and evaluators in the context of a particular country, recognizing the state should not be viewed as the sole source of legitimacy (See Suspitsin in this volume for a version of this analysis). Questions remain about how much significance to give to any one element, particularly considering the potential market influence on higher education. This is perhaps acerbated in the for-profit sector, but is certainly evident in all sectors, and not just in the United States (Clark 1998; Kirp 2003). Gumport (2000) argues that a view of higher education as an industry is the new "dominant legitimizing idea" of the university (p. 68). The implications of this for legitimization of the private sector have not yet been fully explored. As noted above, legitimacy will have to adjust to accommodate the new strength of consumers over professionals in determining what constitutes a legitimate institution of higher education. That, combined with a declining role of the state, indicate that the sources of legitimacy for the private sector are unsettled, and much work needs to be done to understand the intersection of legitimacy and sectoral growth.

Notes

1. See chapter two by Slantcheva in this volume for an alternative take on the implications of Suchman's definition for organizational legitimacy.

2. The HEA establishes the scope of federal policy toward higher education in the United States and is reauthorized every five or six years.
3. Institutional accreditation in the United States can be achieved through either a regional or national agency. Until the 1970s, regional accreditation was reserved only for public and private nonprofit institutions, and it continues to be dominated by traditional higher education institutions (Kinser 2005). In either form, institutional accreditation is necessary for participation in federal student aid programs. Because of the for-profit sector's reliance on these federal monies, legitimacy in marketplace as an institution of higher education essentially requires legitimacy through accreditation, though legitimacy as a skill training institution does not.
4. This question has been important for private higher education in many countries but, increasingly, is seen as less relevant as the private sector expands. In the case of United States, academic capitalism critiques of public and private nonprofit institutions suggest that revenue generation is a critical function of much university activity, calling into question the basis for a nonprofit norm for institutions of higher education. Because "profit" is being generated irrespective of sectoral designation, owners in the for-profit sector argue that a double standard is at play.
5. The CCA is treated here as the "voice" of the for-profit sector. Obviously there may be other voices that contradict the statements of the CCA, and other sources of information about the intentions of the for-profit sector or one of its many independent actors. The shareholder information provided by publicly-traded corporate owners would be one such alternative source. Since the legislative process is the focus of this chapter, CCA lobbying surrounding the passage of the HEA is more relevant for the present purpose.
6. Much of the for-profit sector agenda was included in drafts of the HEA, but for various reasons, work on the Act was suspended for most of 2005. In 2006, legislation outside of the HEA included a provision that eliminated the 50 percent rule, while several other provisions remain to be negotiated in the final language of the HEA that is still before Congress.

Bibliography

Altbach, P.G. 2001. The Rise of the Pseudouniversity. *International Higher Education*, Vol. 25, Fall. pp. 2–3.

Altbach, P.G. 2003. Centers and peripheries in the academic profession: The special challenges of developing countries. In P. G. Altbach (ed.), *The Decline of the Guru*. New York: Palgrave MacMillan. pp. 1–21.

Bailey, T., Badway, N., and Gumport, P.J. 2003. *For-Profit Higher Education and Community Colleges*. Stanford, CA: Stanford University, National Center for Postsecondary Improvement.

Bloland, H.G. 2001. *Creating the Council for Higher Education Accreditation CHEA*. Phoenix, AZ: Oryx Press.

Blumenstyk, G. 2004, May 14. For-Profit Colleges Face New Scrutiny. *Chronicle of Higher Education*. p. A1.

Burd, S. 2004, July 30. Selling out higher education policy? *Chronicle of Higher Education*. pp. A16–19.

Career College Association. 2004a. Action Alert: CCA's "Top 4" HEA issues in new reauthorization bill. Washington, DC: Career College Association.

——— 2004b. Annual report: 2003–2004. Washington, DC: Career College Association.

——— 2004c. Legislative agenda for reauthorization of the Higher Education Act HEA. Washington, DC: Career College Association.

Clark, B. 1998. *Creating entrepreneurial universities: Organizational pathways of transformation*. New York: Pergamon Press.

DiMaggio, P. and Powell, W. 1983. The Iron Cage Revisited: Institutional Isomorphism and Collective Rationality in Organizational Fields. *American Sociological Review*, Vol. 48. pp. 147–160.

Elman, S.E., Beno, B., Crow, S., Morse, J., Rogers, J. R., and Wolff, R. 2004. Letter to Chairman John Boehner and Chairman Howard "Buck" McKeon. Redmond, WA: Northwest Commission on Colleges and Universities.

Farrell, E.F. 2003, February 14. Phoenix's Unusual Way of Crafting Courses. *Chronicle of Higher Education*, pp. A10–A12.

Gumport, P.J. 2000. Academic restructuring: Organizational change and institutional imperatives. *Higher Education*, Vol. 39. pp. 67–91.

Kelly, K.F. 2001. *Meeting the Needs and Making Profits: The Rise of the For-profit Degree-granting Institutions*. Denver, CO: Education Commission of the States.

Kerr, C. 2001. *Uses of the University*. 5th ed. Cambridge, MA: Harvard University Press.

Kinser, K. 2001, November 17. Faculty at the University of Phoenix: A Study of Non-Traditional Roles. Paper presented at the Association for the Study of Higher Education Annual Meeting, Richmond, VA.

Kinser, K. 2005. "A Profile of Regionally Accredited For-Profit Institutions of Higher Education." In B. Pusser (ed.), *Arenas of Entrepreneurship: Where Nonprofit and For-Profit Institutions Compete*. San Francisco: Jossey-Bass. pp. 69–84.

Kinser, K., and Levy, D.C. 2006. "For-Profit Higher Education: U.S. Tendencies, International Echoes." In J. Forest and P. Altbach (eds.), *The International Handbook of Higher Education*. Dordrecht, the Netherlands, and London, UK: Springer Publishers.

Kirp, D.L. 2003. *Shakespeare, Einstein, and the bottom line: The marketing of higher education*. Cambridge, MA: Harvard University Press.

Kraatz, M.S., and Zajac, E.J. 1996. Exploring the limits of new institutionalism: The causes and consequences of illegitimate organizational change. *American Sociological Review*, Vol. 615. pp. 812–836.

Levy, D.C. 2004. *New Institutionalism: Mismatches with Private Higher Education's Global Growth* Working Paper #3. Albany, NY: Program for Research on Private Higher Education, University at Albany, SUNY.

Lovell, C.D. 1997. "State Postsecondary Review Entities: One step forward and two steps back in state-federal relations?" In L. F. Goodchild, C. D. Lovell, E. R. Hines and J.I. Gill (eds.), *Public Policy and Higher Education: ASHE Reader Series* Needham Heights, MA: Simon and Schuster. pp. 338–344.

Miller, J.W., and Hamilton, W.J. 1964. *The Independent Business School in American Education*. New York: McGraw-Hill.

Ortmann, A. 2001. Capital Romance: Why Wall Street Fell in Love With Higher Education. *Education Economics*, Vol. 93. pp. 293–311.

Ruch, R.S. 2001. *Higher Ed, Inc.: The rise of the for-profit university.* Baltimore, MD: Johns Hopkins University Press.
Shea, C. 1998, July 3. Visionary or 'Operator'? Jorge Klor de Alva and His Unusual Intellectual Journey. *Chronicle of Higher Education.* p. A8.
Slaughter, S., and Rhoades, G. 2004. *Academic Capitalism and the New Economy: Markets, State, and Higher Education.* Baltimore, MD: Johns Hopkins University Press.
Sperling, J., and Tucker, R. W. 1997. *For-profit Higher Education: Developing a World-class Workforce.* New Brunswick, NJ: Transaction.
Suchman, M.C. 1995. Managing Legitimacy: Strategic and Institutional Approaches. *Academy of Management Review*, Vol. 203. pp. 571–610.
Ward, D. 2004. *Letter to The Honorable John A. Boehner and The Honorable Howard "Buck" McKeon.* Washington, DC: American Council on Education.

Part III
Concluding Reflections

Chapter Thirteen

Legitimacy and Privateness: Central and Eastern European Private Higher Education in Global Context

Daniel C. Levy

Introduction

This chapter explores how regional findings on Central and Eastern Europe's private higher education legitimacy look in global perspective. The theme is two-fold. On one side, the region's private higher education has in many respects functioned under a cloud of dubious legitimacy. On the other side, it has attained important and sometimes distinctive and multiple forms of legitimacy. These twin statements are comparative, in keeping with the chapter's purpose of highlighting similarities and differences between Central and Eastern Europe and other regions.

Rapid Growth Amid a Lack of Widespread Legitimacy

A Sudden Surge

A common theme in the consideration of private, nonprofit institutions is a legitimacy challenge. The challenge is often intense where a policy field has long been dominated by public institutions. This is the case for higher education in most of the world outside the United States.

Rapid growth amid weak conventional legitimacy (a term suggesting widespread and established trust and support) is something seen globally in private higher education (Levy 2006b). But it is particularly intense in Central and Eastern Europe, in large part because the region's private higher education emergence was unusually late alongside a long standing, well established, and ample, estimable public tradition. The two regions where the emergence and development of private higher education is even later, Africa and the Gulf-states, do not have the backdrop of the tremendous public university tradition, presence, and legitimacy. So Central and Eastern Europe lies at the extreme for the global generalization that private higher education emergence has been sudden, shocking, and unplanned. Moreover, eruption of private institutions in the region has been very concentrated in time. As several chapters show in this book, creation and growth was largely condensed into the first five years or so of postcommunist rule. What came, came very quickly, contributing to the sense of surprise and bewilderment, making acceptance difficult, just as it makes organizational institutionalization difficult.[1] In comparison, the growth of Asian and Latin American private sectors has been more spread out over years and decades, affording more opportunity for surprise to evolve into routine and increased acceptance. Moreover, these regions had clearer institutional precursors for their most recent and intense period of private higher education growth. And as some private institutions took hold, the ensuing mass of other private institutions had precedent. In Central and Eastern Europe, by contrast, there was much more of a meteoric leap in private-institution enrollment from near zero to up to 30 percent of total national enrollment. That, more than the sheer size of present share of total enrollment, is what stands out for the region. After all, many Asian and Latin American countries have an enrollment share in private institutions well above the 30 percent maximum seen in Central and Eastern Europe.

The growth of private sectors in the region was rapid largely because it was comparatively *easy*, fueled by sudden political and economic change, and because enrollments had been remarkably low. Thus, there was opportunity for quick expansion of the private sector in the region, to medium size by international standards. This suggests a contrast between growth (as well as other successes) and conventional social legitimacy: contextual conditions may facilitate new institutional growth and achievement even while much of the public, including the higher education establishment, casts a wary eye.

Intraregional variation in growth can be associated with variation in legitimacy. In some ways, a comparatively large private sector (e.g., Poland) may reflect and build legitimacy. Constituencies expand. The existence of the private sector becomes less unusual and thus eventually less strange. On the other hand, countries with large private higher education sectors (e.g., Georgia, Romania, and Estonia, as well as Poland) are those where proliferation in the private sector is most extreme—including institutions of markedly low credibility. In such cases, large size may exacerbate broad public concern about legitimacy (e.g., Romania).

Unknown and Different

Shocking newness and deviation from established norms naturally make legitimacy problematic. The weak legitimacy of the region's sudden private higher education has much to do with the previously unknown nature of private higher education. Not only was private higher education rarely known in Central and Eastern Europe, it was not well known or established in the region of obviously greatest potential to legitimize Western Europe. Indeed, Western Europe was and remains the world's major region with limited private higher education sectors.[2] As Kwiek's chapter shows, the lack of legitimacy for such sectors helps explain why Europe's Bologna process would largely ignore them, in turn threatening to undermine legitimacy further. Both in reality and even more in myth, Western European higher education was, at its legitimacy peak, epitomized by the ample and prestigious public university, a contrast with U.S. reality.[3] As we will shortly see, most of the new Eastern and Central European private institutions have been fundamentally distinct from a classic university model. They violate what most people have long accepted and respected. The lack of private higher education in most of Western Europe has helped make the phenomenon seem strange to many in Central and Eastern Europe.[4] In fact, it is this contrast between the two European regions that contributes to making the Central and Eastern region seem so striking in the private sector, even though the region's share of total enrollment in the private education sector is not high in the global context. Furthermore, neither in Western Europe nor Eastern Europe did public universities greatly lose legitimacy or collapse on anything like the order seen in Africa and Latin America. Comparatively speaking, they retained legitimacy—and surrendered less space within which new private forms might be seen as legitimately needed alternatives. The point is most apt where university legitimacy is associated with academic quality. African and Latin American public universities are associated with crises of quality whereas in Europe public universities remain associated with quality.

The lack of legitimizing forces has gone well beyond just higher education. Owing to communism, Central and Eastern Europe did not parallel their Western neighbors in having private sectors of secondary and primary education.[5] Most broadly, a culture of planning, obviously intensified under communism, was incompatible with private initiatives. In sharp contrast, in a country like Russia, a statist tradition left little room for a private economy or even an autonomous society. Even short of extreme statist contexts, legitimacy has been associated with a heavy notion of a public interest, with the idea that private interests are divisive, conflictual, and lesser.[6] In fact, as in India and many other countries , "private" has often connotated profit and particularistic self-interest. This is especially so where nonprofits are commercial, a characterization apt for many Central and East European private higher education institutions. Thus, many in Central and Eastern Europe see these institutions as for-profit (Lewis, Hendel, and Demyanchuk 2003), even though only a few countries in the region have legally for-profit private institutions. The connotations of "private" as illegitimate help explain the use of the fuzzy substitute terms "nonstate" and "nonpublic," as the chapters on Poland and Russia show.[7] Where legitimacy is gauged by generalized perceptions, private institutions targeted to particular interests are handicapped.

Similarly, the term and concept, "nonprofit" lacks the historical standing and understanding it has in the United States.[8] Where nonprofit is a viable notion, it often draws on the legitimacy of pursuing public interests, with goals similar to those of public institutions, albeit more with private money and governance. The lack of a viable nonprofit concept sustains the tendency to regard private as for-profit (Powell 2006).

In other political respects as well, Central and Eastern European private higher education institutions have had legitimacy handicaps. As pointed out in several chapters in this book, the state has often been regarded as the pinnacle source for legitimacy, and Neave's chapter shows that private alternatives such as the church often became weak, even outside the Communist settings. In the Central and Eastern European region, the role of state from the beginning of the growth of private sector was very limited, with little by way of money or regulations. So there were few signs of state-certified legitimacy. On the other hand, private higher education institutions often derived their strength from institutions that themselves did not enjoy general acceptance. These include markets or religious and ethnic minority groupings. They may also include foreign connections. We will see later in the chapter how internationalizing forces open fresh sources of legitimacy, but it is also true that there is suspicion of things foreign, especially when associated with privatization and other dynamics popularly linked to social decline. The rise of ultranationalism in countries like Russia can place a stigma on any institution linked (by the institution itself or by public perception) to the West. Another political difficulty, less commented on in the chapters, relates to the hierarchical governance structure of most private higher education institutions. This is the case around much of the world, and it goes against a cultural orientation that there ought to be some kind of collective management or internal democracy.

Our characterization and understanding of obstacles to private higher education legitimacy fit with leading, broad scholarly treatments of the political economy of the postcommunist era in general. Unraveling of the old order had been fundamentally unanticipated so that groundwork was not laid for building the legitimacy of a new order (Ekiert and Hanson 2003). Private is widely viewed as excess individualism and excess markets, both associated with "a severe decline of public virtues" and an "anti-public spirit" (Poznanski 2000, pp. 216, 218). Privatization has often been viewed as a selfish serving of ex-communists and corrupt businesses, contributing to a deterioration of the overall social situation and a rising national identity crisis (Holmes 1997, pp. 329–341). Surveys in the mid-1990s in much of the region showed a lower belief in the contemporary than in the old economic order, the new often associated with aggressive pushing by Western agencies, and showed a loss of hope, trust, and confidence (Nelson 1999, pp. 119–120). So great was the disaffection that some electorates returned communists to power, while ultranationalist, viscerally antiprivate movements gained menacing backing in countries such as Russia (Holmes 1997, pp. 312, 329).

Preponderant Types of Private Higher Education

Organizational legitimacy can depend heavily on the type of organization in question. Among a range of extant private higher education types (Levy 1986), Central and Eastern Europe has spawned mostly the types that are most dubious in legitimacy, with little of the types that typically have higher legitimacy.

Academic elite private higher education is rare everywhere outside the United States. Analysis of a compilation of the world's top 200 universities shows only six private universities not in the United States (http://www.albany.edu/dept/eaps/prophe/data/WorldUniversityRanking2004_ModifiedFromTHES.pdf). Eastern and Central Europe has no such private institution; more importantly, it probably has quite few that could fit into an even much-expanded elite or near elite category. For example, Suspitsin's chapter indicates that perhaps only three to five Russian private institutions can be called, by any stretch, internationally excellent. Reisz's chapter indicates that Romania's private institutions closest to elite standing label themselves comparable to public counterparts, not better, whereas superiority is a claim in much of Latin America and in other places. Another sort of claim that is more modest than elite institutional status is leadership in a niche, a set of undertakings. Business-oriented study is a salient example globally and so it is in Central and Eastern Europe. But the region's niches seem to be fewer than in many other regions and, as the business case illustrates, linked to fields other than those with high academic legitimacy.

A historically prominent type of private higher education in other regions has been religious. This has often meant Catholic, though variety has expanded greatly in recent decades. For various reasons, religious universities have often been seen to be at the academic top of the private sector.[9] As Brazilian, Chilean, Philippine, South African and many other cases show, the religious institutions tend to be older, larger, less commercial, and more clearly nonprofit than other private higher education institutions are. Central and Eastern Europe has not been without religious universities; some predated communism, a few (as in Hungary and Poland) represented the only quasi-private universities under communism, and some have emerged in the postcommunist era. But the examples have been comparatively few, particularly if we focus on religious institutions with major academic legitimacy.

Religion sometimes overlaps with ethnicity, but ethnicity (often including language distinctiveness) alone is the driving force for some private higher education institutions. This international statement fits Central and Eastern Europe. In general, the ethnic group in question is a minority group and the private institution has a role of protecting, sustaining, or promoting the group identity. Such a role can make the institution legitimate in the eyes of that group, even among those who individually do not avail themselves of the private option. Yet others from the same group may themselves deny legitimacy to "their" institution, preferring that the group adopt a more integrative stance within the country's mainstream institutions. Of course, the denial of legitimacy to an ethnic institution comes mostly from other groups and/or from the majority population . Where institutions cater to a particular group, they may not be legitimate or accepted by others. Academic and social legitimacy may be especially problematic if the minority group is seen as low status but, on the other hand, the minority group is often high status, and this may make its institutions politically vulnerable, as some cases in Asia show. The South East European University in Tetovo, Macedonia (functioning from 2001/2002) serves the Albanian minority and spawned confrontational disputes at the founding stage. Ethnic strife has also occurred in countries such as Macedonia and Romania.[10]

But none of these forms (elite, religious, or ethnic) is nearly the most prevalent form in contemporary private higher education either globally or regionally. Instead, it is the commercial type that is numerically dominant. It is this type that most accounts for the unprecedented global boom in private higher education. Regionally and globally, the commercial dominance is particularly dramatic with regard to proportions of institutions, but the point strongly holds even with regard to proportions of enrollment. These points hold both in places where private higher education is mostly a new phenomenon, as in Central and Eastern Europe, and places where a private higher education sector is longstanding (Levy 2006b). Many U.S. colleges that could once be categorized as liberal arts or religious are now more and more basically commercial, whereas they try to maintain their traditional myths for the sake of legitimacy (Delucchi 1997; Breneman 1994). In fact, one of the general contemporary problems for the legitimacy of nonprofit organizations, whether in education or not, is when and how much they commercialize. Financial viability is often jeopardized when they do not sufficiently commercialize, but commercialization undermines the legitimacy claim to being private (nonprofit) institutions in the public interest.[11]

The commercial institutions as a group tend to be farthest from higher education's legitimizing pinnacle, both in mission and in perceived quality. They are often very small, and narrow. Much more than in the public sector, they tend to be "nonuniversities," lacking the grand, legitimizing aura of "university." Often they want to label themselves universities, in large part to build legitimacy and garner the resources that can flow from legitimacy, yet sometimes are blocked by public policies such as accreditation. Sometimes, in fact, the doubt is less about whether they are university or nonuniversity than about whether they are "higher education" or not. Again, all this is common globally, but the region shows rather few exceptions.

Rising Legitimacy

To this point, the chapter has dealt with the fragile legitimacy of Central and Eastern European private higher education. It has focused on the sector's lack of conventional legitimacy. It has highlighted the sector's sudden surge, its largely unknown nature, and its concentration in the least accepted types of private higher education, mostly commercial. Yet some, perhaps all, of these dynamics tend to weaken. This is most clear and obvious regarding the surge and the unknown nature of private higher education. The flip side of the this surge being so concentrated in time is that the share of total private higher education has not ballooned since the mid-1990s. As time passes, what was shocking for newness is no longer so new. A deep breath is possible. There is realization that systems that are 10 or even 30 percent private are not on the verge of becoming 50, 70, or 90 percent private, as has often happened outside the region. Many people still argue that private higher education is strange, unusual, and unworthy, but many get used to it even if grudgingly. Short of approval, a kind of acceptance and recognition provides some legitimacy.

While the passage of time has helped legitimacy, other factors have contributed as well. We proceed to analyze pivotal changes on the public side and then the growth of multiple legitimacies.[12] Although such developments may ultimately prove to be partly reversible, they are important to date.

Public Changes

There is a rising sense of legitimacy for private institutions or at least the opportunity for it, in part because the public sector is changing. Weak private legitimacy has stemmed largely from not being like the much more legitimate public counterpart, hence public internal changes that make the private sector seem less unusual can at least indirectly shore up private institutional legitimacy. We look first at changes within public higher education and then at changes in the state. Neither of these changes is undertaken in order to bolster private higher education legitimacy, of course, but they can have that effect. Additionally, the public side also undertakes public-private initiatives that more directly bolsters private legitimacy.

A major global change is that public higher education decreasingly represents a monopoly. Importantly, this is true for Western Europe. Although private initiatives in countries such as Germany (see Giesecke's chapter) and the United Kingdom are high-profile breakthroughs, the largest privatization in Western European higher education (since the inception of the Central and Eastern European private surge) lies in the partial privatization within public universities. Finance is the clearest example but management and even mission show a shift to the market. As all this translates into increased demands for a posteriori accountability over a priori assumption and trust that the public university will perform well (Neave 1998), it moves public institutions to pressures somewhat more akin to those natural in private sectors. If public places allot an increasing share of admissions slots to private paying students, as they do in some countries in the region and beyond (e.g., Egypt and Kenya), then private institutions charging tuition do not seem so strange and illegitimate. Overall, it may seem a reach for public university personnel and their supporters to decry privateness as inappropriate and illegitimate for higher education.

Even where no privatization occurs, public institutional differentiation can have a legitimizing impact on the private sector. Such differentiation of course predates 1989, and in fact there are de-differentiating tendencies as well (Meek et al. 1996). But clearly "real" higher education is not just an elite university enterprise that involves very serious international standards of research. The point holds regarding the public portrait from the West. More importantly, it holds within Central and Eastern Europe itself. And because pre-1989 cohort enrollments had been so low, both growth and institutional differentiation on the public side surely make such tendencies on the private side seem less extraordinary than they would if they occurred only in the shadow of elite, esteemed universities. Finally, much more in Central and Eastern than in Western Europe, some public institutions are widely seen as corrupt, as with Georgia's admissions process.[13]

No less important than changes in public higher education are changes in the state. The state loses much of its former dominance, especially compared to the communist experience. It lacks the money to finance the post-1989 enrollment boom and it certainly lacks the political control it had over higher education. It depends on private higher education in many cases in order to provide the access to meet the demand which, if not met, would undermine the state politically. Nobody expects the state to resume its monopoly role in finance or rule—or its once unchallenged role as legitimizer. To be sure, the state's legitimizing role was already weakened where the state itself was losing legitimacy. Until a viable, new, postcommunist state takes hold, association with the delegitimized state could hurt an institution's own legitimacy in some respects. Political regime transition has not been a simple challenge, or one universally seen as successful. Certainly, the Central and Eastern European region and the countries previously part of the USSR are not all legitimate democracies; some are rather outright dictatorships and others are "semi-authoritarian" regimes (Ottaway 2003).

However much the changed state or public higher education sector makes private higher education legitimacy easier to obtain, the public actors—both state and higher education—also bolster the legitimacy through direct action with private institutions. This occurs in several ways.

Only exceptionally, Central and Eastern European states give direct subsidies to private higher education but, as in Poland, more comes through indirect help, perhaps via students, or through competitive research funding. As in Hungary, governments may contract private institutions for particular activities. Postcommunist municipal governments often donated land and buildings. Then there is the impact on private institutional legitimacy where the public sector pays full-time salaries to its professors who give additional classes in private universities, and most private institutions have many more part-time than full-time professors. The sharing of professors may be recognized or just accepted de facto, even where it contradicts official policy. In any case, it is a tangible support by both state and public university for the private university. Thus, it is at least an indirect and partial legitimizer. Public university professors bring some instant legitimation to the private sector, as they carry trust, prestige, status, pedagogical methods, and curriculum. All this enhances organizations previously weak in legitimacy, as it allows for "professional isomorphism" and "normative isomorphism" (Levy 2006a). And all the points in this paragraph follow global patterns.

Regulations and accreditation also represent powerful legitimizing forces from both the state and public universities. Although the initial lack of regulations for the surging new private sectors was important for the surge to be possible, it also denied an official stamp of approval. Licensing and "delayed regulation" (Levy 2006b) assures many that at least minimum rules are in place. Accreditation, widespread on the global higher education reform agenda, gives more of an official stamp. Even where accreditation is granted by government or a related national body, accreditation personnel, modes, and standards often come largely through public university professors.

Obviously, increasing regulatory stringency can be the ultimate delegitimizer for private institutions that cannot make the grade.[14] Even if allowed to operate, they are flying on one engine. Fully accredited institutions, however, have extra state fuel in their tank.

Perhaps the most dramatic legitimizing collaboration comes through formal private-public partnerships. Such partnerships are on the rise internationally, and not just in higher education (Salamon 1995). They typically involve mutual interest based on a linking of each side's comparative advantages. In higher education the private side may bring tuition-paying students, supplemental income for public university professors, nonuniversity training, job market orientations, and distance or other low-cost offerings. The public side may provide costly and advanced facilities, academic expertise, quality, status, tradition, trust—and legitimacy. Most of South Africa's private higher education institutions are engaged in formal partnerships with the country's public universities. China is zealously experimenting with different match ups. Yet few cases abroad show the intensity seen in Russia (see Suspitsin's chapter), including the public university role in creating private institutions.[15]

Public parties to partnerships may or may not have a principal aim of bolstering their private partners' legitimacy. They may be in partnership simply for their own benefit. But they can hardly be unaware of the legitimizing effect. Even if some of their professors or students remain among the harshest critics of private institutions, it becomes harder to turn around and say that these private institutions are not legitimate.

Rising Legitimacies: The Many over The Central

So legitimacy increases from an array of public and public-private dynamics. We now look at how these increases also stem from the rise of multiple legitimacies. This rise is intertwined with a shift in the nature or perception of legitimacy itself. A conventional sense of legitimacy connected to society at large, the state broadly, and Suchman's (1995) notion of a generalized perception, yields some ground when we consider multiple sources and types of legitimacy.

A concept of multiple sources and types of legitimacy has a logical fit with private sectors more than public ones, especially where public systems are nationally centralized (as in the "Continental model"). (Where public systems substantially privatize, multiplicity becomes more likely than before.) Private sectors often pursue private interests or private and public interests more than "the public interest." It is not that individual private institutions are more plural than are public counterparts. On the average, they probably have a narrower legitimacy focus and may not pursue multiple legitimacies. The notion of multiple legitimacies applies more at the sectoral level. Different private institutions may pursue different missions and different ways of being accepted by different populations. Private institutions usually need not try to appeal to and gain acceptance from all. People can choose among alternatives rather than be bound to one paramount norm and model.

We may consider multiple legitimacies as, often individually and rather narrow yet, sometimes quite deep, that is, a given institution may not have to gain legitimacy from the public at large as much as from some particular actor or a small set of actors; however, this legitimacy might have to reach levels of activity and participation beyond just vague acceptance. A common example is willingness to pay the institution's tuition or to hire its graduates. In fact, private higher education institutions can sometimes exist and even flourish in the face of a lack of broad popularity and legitimacy, even with some wide view that the institutions are illegitimate. Unless those perceptions translate into proscriptions or heavy restrictions, private institutions may exist or even thrive based on just the deep support they get from some. "Niche legitimacy" can be viable. There can be different routes to legitimacy and even different types of legitimacy.[16] Different groups or constituencies may evaluate different private higher education differently. They may not accord great legitimacy to the sector as a whole while according legitimacy to an institution or group of institutions. Or they may accord legitimacy to the sector, while denying it to certain institutions within the sector.[17]

Religious and ethnic focused institutions are often clear examples of niche-based legitimacy. Baltic countries present examples, as with the Russian population in Estonia. Just as we noted that ethnic tensions may mean that private institutions lack legitimacy with the majority population, such tensions may reinforce strong attachment by a group to its own institution. Where there are no such tensions, notions of pluralist tolerance may gain strength, thus in turn bolstering the legitimacy of institutions with special target populations.

The growth of plural civil society can be as conducive to multiple legitimacies as the communist repressive state was hostile to it. Equally important regarding the context within which higher education lives is the emergence of market economies. Notwithstanding the tendency (a delegitimizing tendency) for there to be huge concentrations of power and wealth, competition and multiplicity grows, certainly compared to the communist era.

The government also becomes more plural or decentralized. We have noted the role of local governments in helping private institutions. Additionally, as in Poland and Hungary, the private institutions often have a locally based legitimacy. As in other parts of the world too, national public universities may look down on such local institutions, while local communities and employers may welcome the status, access, and job training they provide.

In China, starting in the 1980s, the central government coyly monitored the varied activity of local governments. They watched private institutions grow. They did not comment too much. Though there was always the danger that they would suddenly denounce private and local initiatives, they did it only sporadically. They observed performance, decided that private institutions were indispensable to mammoth expansion, and probably relished teaching public universities a few lessons about innovative aspects of privatization. And then they blessed the private higher education institutions with legitimizing legislation in 2002. But even before that, the authoritarian central government left the local governments and provinces with considerable (though fragile) autonomy. Less startling is that democratic India allows very varied policies and private realities in different states.

Thus, multiplicity of missions and legitimacy sources may be promoted by great diversity of local or state government policy. The United States has long been the most important example of interstate policy diversity. Some U.S. states are more favorable than others in the recognition and treatment of private higher education (Zumeta 1992). In other countries, some states or localities remain hostile to private initiative, while others are more enthusiastic than the national government. Legitimacy grows in some places even as it is lacking in others.

Additionally, even at the national government level (in Central and Eastern Europe) there is a diminished sense of the unitary and an increased reality of the plural.[18] The presidency and the ministry retain power, but other branches of government have a place too. Courts play a role in interpreting and enforcing central law. Yet it is mostly legislatures that become serious policymaking bodies—often the leading policymaking bodies, a striking change for the region. Echoes are heard in local government and its legislature but most countries in the region remain national in their policymaking. Whether local or national, legislatures tend to be venues at which different groups and interests—including those closely tied to particular private institutions—gain representation, influence, and legitimacy. In turn, enrollment growth can enhance such political weight. All this finds ample precedent from Latin America.

For all the diversification of actors and sources of legitimacy within countries, there is a potent internationalization as well. If nation-state-granted legitimacy has an elective affinity with public higher education, international legitimacy does not. Globalization fits with the remarks made above, about seeking legitimacy through expanding marketplaces. Regional (European) economic and political forces and institutions also figure here, though Kweik's chapter on the Bologna process reminds us that not all additional or rather non-national forces need to promote private higher education legitimacy. Still, private institutions suffering a legitimacy deficit at home can look hopefully for support from abroad. Sometimes no international institutional engagement is required, as when private institutions simply call themselves "international" or "American." A more energetic route is accreditation with a foreign agency. Others include student or faculty exchanges. Still others may come with formal partnerships, which may follow a logic overlapping of domestic private-public partnerships. In this case, however, the status and legitimacy come mostly from abroad, from universities there seeking wider markets and income. Private institutions in Central and Eastern Europe can look to universities in Western Europe. The United States is another common legitimizing source. This is not to ignore that international identification can be a curse as well as a blessing, as noted in the case of Russian ultranationalism.

Private Institutions' Efforts Toward Legitimation

Private institutions that are not just beneficiaries of changing forces may themselves try to build legitimacy, in the evolving environment, as several chapters in this book show. Some seek legitimacy through pointed emulation of public norms. Others pursue distinctiveness. Many pursue a combination, with great variation in the mixes.

Choices and strategies are involved but equally important is deftly taking the opportunities that come one's way. The region's private institutions can hardly opt to become high- status research universities. They can find roles connected to the new economy. In a context of soaring demand for higher education and limited public space and funds, they can stake a major claim to legitimacy through access (though persistence of high demand exceeding public supply is not assured in Central and Eastern Europe, as it is in India, China, and other developing regions). The legitimacy claim of offering increased opportunity strengthens if the institutions can achieve a decent level of academic quality and not appear to be merely commercial demand-absorbers. Beyond that, a minority of institutions attempts to claw its way higher than that, albeit not to the academic peak of the region's national systems; they aspire for semi-elite or mid-range status, perhaps with a few niches of excellence. Such goals help explain attempts to hire more full-time faculty or increase the facilities and offerings. None of this is unique to Central and Eastern Europe.

Nor is the region exceptional for explicit private sector claims to performance-based legitimacy. Of course we want to know how and to what degree private institutions perform, but the point here is that performance is now a major basis on which to achieve legitimacy. The new political economy welcomes (even demands) this sort of accountability. Focus on "measurables," rather than myths, reputations, and trust, again goes against the classical Continental model where legitimacy is supposed to be a-priori—you are there, you are an arm of the state, you are assumed to be doing what is legitimate and what is good, often in a leadership role. In today's era of heightened performance demands, it may be more about serving and fitting outside interests than about leading. In fact how much public institutions need to be responsively "relevant" or suffer an erosion of legitimacy is unclear. Private institutions, at least except for diploma-mills, might have to face the new pressures but can boost their legitimacy if they do so successfully. Unable to draw on a deep well of traditional legitimacy, their only viable route to legitimacy may run through the market.[19] "Externalistic" and "output-oriented" are terms used in Reisz's chapter and "pragmatic legitimacy" is a term used by Suchman (1995). This fits very well with the private higher education literature's decades-old finding that private institutions concentrate in fields of practical job market orientation. Central and Eastern Europe further confirm that empirically-based generalization.

Private higher education institutions often claim to highlight student-centered satisfaction and even focus laser-like on serving student interests. In addition to access, these interests concern jobs and practicality. The claim has to be even sharper when it comes to for-profit institutions. U.S. for-profits such as those analyzed in Kinser's chapter point to surveys showing high satisfaction. Those surveyed include graduates who refer to performance-based results. The U.S. findings are pertinent since they show the largest database and since many commercial nonprofits in Central and Eastern Europe and beyond share some salient characteristics with U.S. for-profits institutions.

Conclusion

Most of what we find regarding private higher education and legitimacy in Central and Eastern Europe is found elsewhere as well. It is largely the degrees and particulars that are different.

Sudden private higher education growth is common globally but was especially intense and stunningly concentrated in Central and Eastern Europe. Yet from the outset, an array of potent forces would undermine aspirations for ample private sector legitimacy. These forces included statism, nationalism, a norm of unified systems and centralism, and a commitment to pursuing *the* public interest. In contrast, diverse societal initiatives, decentralization, nonprofit sectors, and honoring of private interests had long lacked expression and legitimacy. Though the capacity of the communist government to legitimize crashed, there has been a persistent legacy of old forms and beliefs. A further large problem has been the concentration of private sector growth in academically weak institutions, mostly commercial.

Nonetheless, nowhere more dramatically than in Central and Eastern Europe, changes in domestic as well as international politics and economics have opened fresh routes toward private institutional legitimacy. The state weakens as provider of resources and legitimacy, even as it remains the single most important actor on both scores. Some public higher education institutions suffer legitimacy deficits themselves. At the same time, public higher education diversifies and even partly privatizes. New forms of finance, governance, and accountability make private sector practices and norms seem less radically or unacceptably out of the mainstream. There is increased need to develop legitimacy through market and international spheres and many private (and public) institutions see opportunity as much as challenge there.

The notion of narrow yet adequate legitimacy from particular constituencies becomes more pertinent. Different institutions are accountable to different actors and gain legitimacy from different ones. The growth of markets, civil society, decentralized government, and various globalizing forces increases the number of potentially legitimizing sources and the idea that different institutions can build different trajectories and legitimacies. Private higher education can seek legitimacy through different kinds of performance-based legitimacy.

None of this provides strong and secure legitimacy to the region's private higher education overall. Norms and practices do not fully change overnight, revolution notwithstanding. Private institutions are but the young siblings, still vulnerable to serious and even disabling legitimacy problems. Nor should we slip into assuming a linear road from weak legitimacy to more and more legitimacy. Thus the picture on legitimacy is mixed, fragility alongside notable gains. Such a mix is common internationally, though both the weaknesses and the gains are particularly striking in the dramatic and still young postcommunist region.

Notes

1. Literature in political sociology has noted something of an inverse relationship between growth and institutional viability with credibility. Slower growth may allow time for institutions to develop their norms. See Huntington (1968).
2. Aberrant Portugal has an ample private sector (Teixeira 2004), 26 percent private (comparable to the upper end of that seen in Central and Eastern Europe) and many countries have long had what Geiger (1986) aptly calls "peripheral" private sectors, modest in size, often religious or otherwise specialized. That leaves only Belgium and the Netherlands with large and important private sectors, but these have been only ambiguously private as they have been publicly financed and not much different from the public universities.

3. The European public university could focus on training for esteemed professions, research, or both. Private higher education is a very accepted form in the U.S. and surveys there show high scores in legitimacy (Salamon 2003: 31). Whereas U.S. private higher education is much higher in legitimacy than its Central and European counterpart, the share of enrollments is similar, neither particularly high nor particularly low by global reality.
4. Private institutions may attempt to counter seeming strangeness born of newness by copying public university forms and practices. But it is difficult to copy at the same level of quality or resources. And if you do not do something distinctive, then who are you and why should you have a presence?
5. In much of the world outside the United States, private sectors were long established at secondary and primary levels before they emerged at the higher education level.
6. A pertinent literature on such points contrasts "corporatism" with "pluralism," the former more characteristic of European tradition, the latter more characteristic of U.S. tradition. Corporatism can be "state" or "societal," in both cases drawing legitimacy from concepts such as harmony and publicness much more than self-interested pursuits and privateness. See, for example, Schmitter 1974.
7. The non-state label holds not just for postcommunist higher education but also for much primary and secondary education in Western Europe.
8. Salamon 2003. Yet, despite the strong associational propensity, Tocqueville found remarkable about the young US, nonprofits did not always have the widespread legitimacy they now enjoy. They were often denounced as particularistic and contrary to harmony and the common good (Neem 2003, pp. 344–354).
9. Often, their key legitimacy problem stemmed more from society's rejection of the very notion of non-secular universities. Religion in Europe and Latin America was pushed out of the central national universities, those with highest public standing, and permitted only in private niches.
10. A related or parallel inter-group clash deals with regions within a country. This is the case in Ukraine, between Russians and Ukrainians. A group that is a minority in the country may be a majority in a given region.
11. Even leading universities put their legitimacy at risk by intense commercialization (Bok 2003).
12. Just as emergence of private higher education in Central and Eastern Europe shows a certain emulation of U.S. experience, so does the subsequent pursuit of legitimacy. In the earlier part of the twentieth century the U.S. pursuit was largely through building academic credibility (Jencks and Reisman 2001), whereas by the later part of the century it came increasingly about market credibility. If the US had had European orientations toward rule-making, coordination, quality assurance mechanisms and definitions, many of its private institutions a) would never had been born or b) if born would have died. Many U.S. private institutions have died, but mostly through the market. It seems that many countries have moved toward latitude not only in allowing private birth but also space for activity and possible improvement.
13. A large question for transitional countries is what kinds and degrees of corruption are more associated with the public sector on the one hand and the private

sector on the other. One hopeful idea for both sectors is that the growth of market competition restrains corruption because corruption simply cannot be afforded; even if one accepts the argument long-term, it is unclear how much weight it carries near-term.

14. Regulation and accreditation may also bring a more subtle risk to private institutional legitimacy. They may push institutions away from distinctive undertakings without which they lose their raison d'tre. Or they may cripple the institutions in the marketplace, as when they are forced to hold tuition below a certain figure. A risk/reward dynamic also holds for public higher education institutions facing the peak universities.
15. Especially where corruption is rampant, state officials may be involved for private gain more than state policy.
16. Pachuashvili's chapter is quite relevant in respect to the multiplicity of sources and types of legitimacy in postcommunist countries. Nicolescu's chapter reminds us that commitment to private institutions need not imply a deep, value-based identification that these institutions have very high standing; like many regional counterparts, Romanian students usually still see some public universities as the pinnacle, but where they make private universities viable second choices they are treating these universities as quite acceptable. Perhaps allegiance builds during the years of study and thereafter.
17. Multiple sources of legitimacy do not usually mean a great number of authorities over a given institution. As pointed in the Introduction to the book, Meyer and Scott (1983, p. 202) associate multiple controllers with a lack of institutional focus and with a negative impact on an institution's legitimacy, but most private higher education institutions in Central and Eastern Europe, and most of the world, tend to be heavily accountable to a few actors rather than accountable simultaneously to many.
18. Just as the rise of local and state government marks a tendency long characteristic of U.S. higher education, a kind of "division of power," so there is a rising "separation of powers" among branches of government.
19. It is unclear how much which private institutions pursue legitimacy directly or simply pursue goals such as job-oriented training, which then, if successful, build legitimacy. Either way, legitimacy through performance is at least a possibility.

Bibliography

Bok, D. 2003. *Universities in the Marketplace: The Commercialization of Higher Education*. Princeton: Princeton University.

Breneman, D. 1994. *Liberal Arts Colleges: Thriving, Surviving, or Endangered.* Washington, D.C.: Brookings Institution Press.

Delucchi, M. 1997. Liberal Arts Colleges and the Myth of Uniqueness. *The Journal of Higher Education*, Vol. 68, No. 4, pp. 414–426.

Ekiert, G. and Hanson, S.E., 2003. Introduction. In Ekiert, G. and Hanson, S. (eds.), *Capitalism and Democracy in Central and Eastern Europe: Assessing the Legacy of Communist Rule*. New York: Cambridge University Press, pp. 1–14.

Geiger, Roger. 1986. *Private Sectors in Higher Education: Structure, Function, and Change in Eight Countries*. Ann Arbor: University of Michigan Press.

Holmes, Leslie. 1997. *Post-Communism: An Introduction*. Durham: Duke University Press.

Huntington, S. 1968. *Political Order in Changing Societies*. New Haven: Yale University.

Jencks, C. and Riesman, D. 2001. *The Academic Revolution*. New Brunswick: Transaction Publishers.

Levy, D. 1986. *Higher Education and the State in Latin America: Private Challenges to Public Dominance*. Chicago: University of Chicago.

——— 2006a. How Private Higher Education's Growth Challenges the New Institutionalism, in H.D. Meyer and B. Rowan, (eds.), *The New Institutionalism in Education*. SUNY Press. Albany, NY.

——— 2006b. The Unanticipated Explosion: Private Higher Education's Global Surge. *Comparative Education Review*, Vol. 50, May, pp. 217–240.

Lewis, D. R., Hendel, D. & Demyanchuk, A. 2003. Private Higher Education in Transition Countries. Kiev, Ukraine: KM Academia Publishing House. Reprinted in Azeri and Russian at Azerbaijan International University.

Meek, L., Goedegebuure, L., Kivinen, O., and Rinne, R. 1996. *The Mockers and Mocked; Comparative Perspectives on Differentiation, Convergence and Diversity in Higher Education*. New York: Pergamon.

Meyer, J. and Scott, R. 1992. *Organizational Environments: Ritual and Rationality*. Newbury Park, CA: Sage Publications.

Neave, G. 1998. The Evaluative State Reconsidered. *European Journal of Education*, Vol. 33, pp. 265–284.

Neem, J. 2003. Politics and the Origins of the Nonprofit Corporation in Massachusetts and New Hampshire, 1780–1820. *Nonprofit and Voluntary Sector Quarterly*, Vol. 32, pp. 344–354.

Nelson, Daniel. N. 1999. Civil Society Endangered. In Sakwa, R. (ed.), *The Experience of Democratization in Eastern Europe*. New York: St. Martin's Press Inc, pp. 118–137.

Ottaway, M. 2003. *Democracy Challenged: The Rise of Semi-Authoritarianism*. Washington D.C.: Carnegie Endowment for International Peace.

Powell, W. and Steinberg, R. 2006. *The Nonprofit Sector: A Research Handbook* 2d edition. New Haven: Yale University Press.

Poznanski, Kazimierz. Z. 2000. The Morals of Transition: Decline of Public Interest and Runaway Reforms in Eastern Europe. in Antohi, S. and Tismaneanu, V. (eds.), *Between Past and Future: The Revolutions of 1989 and Their Aftermath*. New York: Central European University Press, pp. 216–246.

Salamon, L. 1995. *Partners in Public Service: Government and the Nonprofit Sector in the Modern Welfare State*. Baltimore: Johns Hopkins University Press.

Salamon, L. 2003. *The Resilient Sector: The State of Nonprofit America*. Washington, D.C.: Brookings Institution Press.

Schmitter, P. 1974. Still the Century of Corporatism? in Pike, F. and Stritch, T. (eds.), *The New Corporatism*, Notre Dame: University of Notre Dome Press, pp. 85–131.

Suchman, M.C. 1995. Managing Legitimacy: Strategic and Institutional Approaches. *Academy of Management Review*, Vol. 20, No. 3, pp. 571–610.

Teixeira, P., Rosa, M.J. and Amaral, A. 2004. Is there a Higher Education Market in Portugal? in Teixeira, P., Jongbloed, B., Dill, D., and Amaral, A. (eds.), *Markets in Higher Education: Rhetoric or Reality?* Dordrecht, Boston, and London: Kluwer, pp. 291–310.

Zumeta, W. 1992. State Policies and Private Higher Education. *Journal of Higher Education*, Vol. 63, No. 4, pp. 363–417.

Chapter Fourteen

Reflections on Private Higher Education Tendencies in Central and Eastern Europe

Peter Scott

Introduction

An intriguing and broad question that one can pose while reflecting on material in this book is how much the strong development of private higher education in the countries in Central and Eastern Europe is an exceptional phenomenon or the part of the general evolution of higher education systems toward greater diversity, pluralism, and differentiation (of which the growth of private institutions may simply be one aspect). Is the development of private higher education in the post-communist world better explained in historical or structural terms—in other words as a response to a particular set of historical circumstances (and, by implication, limited by these historical origins); or as a set of structural and organizational adjustments to the demands of contemporary society (which is typically characterized as a "knowledge society" shaped by the forces of globalization?) These are large matters that do not lend themselves to easy or provable answers but the exploration can be illuminating.

Closely related to this primary question is a secondary question that is particularly relevant to the growth of private higher education. While the legitimacy of traditional universities (and other higher professional institutions), and of public systems of higher education, is (was) clearly grounded in notions of the "public good," that can be traced back to ideas of enlightenment, emancipation, opportunity, and progress developed in the nineteenth and twentieth centuries, the legitimacy of private institutions and of market systems is less securely grounded. This relative insecurity is reflected in terms both of governance and funding (the implied triumph of "managerialism" over collegiality, and the controversies about tuition fees and student support) and of the intellectual and scientific life of universities (the emphasis on more open and distributed knowledge production systems, and the threatening ambiguities of post-modernism). Once again, the question arises whether the difficulties that the private higher education institutions in post-communist countries have experienced in establishing their legitimacy can be attributed more to specific historical circumstances—the amalgam of utopianism, opportunism, and cynicism that tends to accompany any regime change—or are better explained in terms of this wider crisis of legitimacy experienced by most post-public higher education systems? In chapter two, Slantcheva directly addresses this question—are the difficulties encountered by private institutions in acquiring legitimacy transitory because they are essentially the product of once dominant social and political norms that are now fading or are they more intractable?

The development of private higher education in Central and Eastern Europe (historical contingency or structural adjustment?), of course, must be seen within the wider context of the development of private higher education in other regions. Here there is substantial variation. In North America, South America and some Asian countries such as Japan and Korea, private institutions have long been prominent in higher education systems; in contrast to Western Europe where private higher education has struggled to acquire any purchase on what has remained predominantly a public system. In the newly emerging giants of global higher education such as China and India, new configurations of public and private provision are emerging. On a higher plane, this variation mirrors the tensions between historical contingency and structural adjustment apparent in post-communist higher education systems in Central and Eastern Europe—but may also expose the variability of definitions, even volatility of language, in assigning private and public labels to institutions within increasingly differentiated higher education systems.

In this chapter the following questions will be explored. The chapter is divided into three sections:

1. The wider context: part one—the impact of the transition from communist to post-communist societies on policy experiments in general and the fortunes of private higher education in particular;
2. The wider context: part two—the emergence of more differentiated higher education systems and more adaptable institutions in response to the growth of a knowledge society, and the extent to which this differentiation and adaptability is promoted by various forms of privatization;
3. The evolution of post-communist higher education systems in Central and Eastern Europe (including the former Soviet Union)—and the origins, extent, impact, and significance of private institutions within that evolution.

The Wider Context—The Transition to Post-Communist Societies

Although it may not be a welcome or comfortable comparison, this author is intrigued by similarities between the first decade of the former Soviet Union and the first post-communist decade. This is not to imply that dynamics were the same in specifics or degree. Both were periods of fundamental experimentation. In the Soviet Union in the 1920s, after centuries of sedimented Tsarist rule, there was an explosion of novelty—most notably in the attempts, both utopian and absolutist, to build a communist state and a communist society but also in the transformation of social relationships and individual identities within this new kind of state and society. This was reflected in a period of creative turmoil in the arts (though, as noted below, Russia of the late nineteenth and early twentieth century was a time of greater cultural and lifestyle innovation, much more than would be the case for the post-communist period). One thinks of Meyerhold in the theater, Tatlin and other constructivists in architecture, and Blok, Gumilev, Mayakovsky, Mandelshtam in literature. In fact, creative forces limped tragically into the Stalin period. In former communist countries in the 1990s, after a half-century more of triumph and terror and (more recently) ossification and stagnation, there was also a period of unconstrained novelty—privatizations on a scale unimaginable in comfortable welfare-state oriented western Europe (or even the United States), the emergence of a new politics contested between the fabulously wealthy oligarchs created by these privatizations and populist-traditionalists who continued to draw their strength from older traditions of state power, a radical questioning of the public because of its tainted association with communist-dominated institutions and mentalities.

This analogy between the 1920s, in the former Soviet Union, and the 1990s, in the wider post-communist domain, should not perhaps be pushed too far. There was little, if any, sense of novelty and experimentation in the countries of Central and Eastern Europe that fell under the sway of the Soviet Union after the defeat of Nazi Germany to match the excitement of the 1920s in Soviet Russia—outside the tight and disciplined ranks of the party cadres (and perhaps, very briefly, among a wider section of progressive nationalists). Stalin was no more successful in winning the hearts and minds of his new subjects to the cause of Communism than Tsar Alexander I had been a century and a half earlier in winning over middle European opinion to the cause of the Holy Alliance. Both failures inevitably led to retreats—the collapse of the Soviet empire (internal and external) in 1989–1991 and the less dramatic but nevertheless decisive displacement of Russia from the heart of the European concert of great powers beginning in 1848–1849 accelerated by the Crimean War and continuing until the collapse of Tsarist rule (There are, incidentally, intriguing parallels between these two processes of retreat—presaging perhaps similar eventualities). Another important difference is that the period of experimentation in the Soviet Union in the 1920s was terminated by a brutal dictatorship while the current round of social and economic experimentation (but curiously not cultural experimentation) in the post-communist world—hopefully—will lead to the establishment of stable democratic states with vibrant civil societies.

But what is common to these two periods is that both succeeded a long period of apparently stable order that had yielded to decay and delegitimation. Although the ultimate unravelings would be rapid, and in a sense unanticipated in both 1917 and 1989–1991, toward the end there was a widespread belief that the regimes were limping along. Few observers of prerevolution Russia thought the regime could survive long and in fact there was considerable political volatility, obviously unmatched in the late communist period. When a long period of stability and a shorter period of obvious decay both decisively ended, they left a strong need and challenge to rebuild notions of legitimacy essential for the functioning of healthy institutions. The problems faced by both the new Soviet regime in Russia in the 1920s and the post-communist regimes across the wider region in the 1990s were similar. Both relied on state and social structures that had previously existed, which was why the early Soviet state (and even party) bureaucracies owed so much to their Tsarist predecessors (and why Tsarist officers fought in such significant numbers with the Red Army) and also why communist bosses in the 1980s so often reappeared as post-communist politicians or even privatization oligarchs. Yet both regimes had to come to terms with the fact that these state and social structures had been created by and for old regimes that had collapsed (and the legitimacy of which had been fundamentally undermined by the harshest judge of all, historical events however contingent). So their open reliance of replication of these structures was both embarrassing and compromising.

This dilemma has been further sharpened because of the sudden shock of what would now be called regime change. As noted above, no one in 1917 had anticipated immediate revolution—which, of course, in retrospect turned out to be a unique event in the context of European if not world history (which, in turn, led to the cruel dilemma of building socialism in one country that so corrupted Stalin's Russia). Moreover recent studies have bolstered the scholarly case that early twentieth-century Russia was not a moribund society, however ossified the Tsarist state. During 1906–1911, economic growth rates exceeded those in much of western Europe and the United States. Cultural achievements were unparalleled; Tolstoy's and Chekhov's Russia like Shakespeare's England or, perhaps,

Goethe's Germany was a place in which human sensibility reached new, almost mythic, heights, which is not to imply that political life showed a remotely parallel greatness. Although it has yet to become so fashionable, a revisionist account of the Soviet Union is gradually emerging—largely, and paradoxically, because of access to Soviet archives that offer a more nuanced and more moderate picture of the realities of life in the Soviet Union than the highly ideologized evil empire totalitarian visions familiar from the cold war period. For example, Moshe Lewin has described a society that in many ways was normal—and that had generated its own legitimacy (although this legitimacy did not extend to most of the communist regimes in Central and Eastern Europe, only the normality) (Lewin 2005).

Hence successor regimes have had to come to terms with social values shaped by the regimes they have replaced—and which continue to be resilient (and not only among the elderly). For example, the new Soviet regime had to come to terms with the persistence of Russian patriotism and nationalism, which is why the nationalities question was one of the most contested political questions in the early 1920s (actually rather more contested than disagreements about policies toward the peasantry and industrialization until the mid-1920s). Similarly, post-communist regimes have had to come to terms not only with traditions of respect for state authority that are excessive by western European standards (but which perhaps have not troubled them greatly) but also with deeply ingrained and often negative attitudes toward private enterprise, often still associated with sub-criminal spivvery and at any rate antisocial individual aggrandizement, a point made in the preceding chapter of this book. However, at the same time, the collapse of the old orders sharply reduced the constraints that often limited radical social and economic experimentation in more gradual periods of transition. For example, legal frameworks developed by the former regimes lost much of their credibility where they did not collapse outright—and were sometimes subject to over-zealous, premature, and ill-considered reforms (a kind of bonfire of state regulations); or, where no or fewer formal changes were made, these frameworks were frequently circumvented leading to a proliferation of gray areas, and sub-legal arenas. There was also a heightened expectation of change; both opponents of the old regimes and opportunists were hungry for reforms.

In considering the development of post-communist higher education systems—and in particular the growth of private institutions—it is good to take into account this wider context. It may help to explain why there were both fewer pragmatic inhibitions about developing private institutions than in western Europe (and fewer obstacles than might exist in the United States in terms of licensing and accreditation) but also greater normative antipathy, and even active resistance, to private higher education (in particular from the public universities—even when they themselves were active playing their own privatization games). The comparison with the dawn of Soviet rule in Russia, itself a period of radical experimentation, is illuminating. Another comparison, equally illuminating perhaps, is with South Africa (Cloete 2002). Although the scale of social and political change arguably has been greater in South Africa than in most post-communist countries, which is why the word "transformation" has been typically used rather than the more modest "economies in transition" label often applied to post-communist regimes, the degree of continuity in higher education structures has been much greater (and continuity as actively willed not as a residual inertia). For example, attitudes to private higher education remained remarkably consistent despite regime change. Part of the reason may be that the process of transformation was nevertheless a negotiated one, shaped by agreed constitutional changes, which meant that key elements of the old apartheid regime continued (and continue) to exist. Part of the reason may also be that the new African National Congress Government was (and is) strongly committed to public, quasi-social democratic values—which in an era when such values are out of fashion may gave the (misleading?) appearance of its being lukewarm toward reform, largely defined in terms of free markets and privatization.

Other comparisons, for example with Latin America, may also be illuminating (Levy 1986). Here, the similarities with the post-communist countries are greater—the extension of private higher education against a background of continuing social disapproval (or, at any rate, sharp political divisions). But there are also important differences: the World Bank, International Monetary Fund, and other global institutions played a more active, but also a more contested, role in the Latin American economy (though not a markedly or decisive pro-private role in the higher education sphere), with the result that privatization could be represented to many as an externally imposed phenomenon (which in turn created crises of legitimacy). The point that should be emphasized is not that crude and reductionist parallels can be drawn between different reform processes in higher education through references to larger political, social, and historical transformations (because these transformations are themselves highly contextualized), but that these transformations both steer and limit the possibilities of higher education reform. The counterpoint between experimentation and continuity/tradition/structure in post-communist countries has its own particular dynamic—as the similar counterpoints in Western Europe (channelled through the Bologna process) or South Africa (the politics of transformation), or Latin America (where the collision between traditional higher education systems and the imperatives of globalization are perhaps most keenly felt) have theirs.

The Wider Context—Differentiation in Higher Education

The development of mass higher education systems, once the object of close and detailed analysis, has become a decidedly less popular topic. One reason no doubt is that the canon of mass higher education, although claiming to offer generalizable accounts, was generated in a particular place—the United States—and at a particular time—roughly speaking, between the 1960s and the 1980s. As a result these accounts tended to privilege one particular driver of massification, growing demand from potential students who regarded participation in higher education as a quasi-democratic right (which chimed rather better with the political culture and social values of the United States than of Europe—and certainly of Central and Eastern Europe). Private higher education barely figured in these accounts, probably because the engines of first-wave massification were state universities and community colleges in America (and similar public institutions in Western Europe); private institutions were sometimes seen as part of the problem, with over-restricted access, rather than the answer. To the extent these accounts of massification addressed issues of diversity and differentiation (which reflected the second big driver, the need to develop more adaptable and flexible systems to accommodate much larger and more heterogeneous student populations with more varied needs and ambitions within the context of an emerging knowledge society) it was firmly within the context of state action. That is why state master plans that carefully stratified institutions by mission received such attention.

More recently, and particularly in the past decade (almost the exact period during which distinctively post-communist higher education systems have emerged), the terms of both academic enquiry and policy formation have changed. The links between systemic reform and institutional variation on the one hand and the market on the other, however imprecisely defined, now receive greater emphasis. This change is not confined to higher education; it also extends to the wider research system where more open and distributed knowledge production systems are increasingly emphasized. There are two reasons for this change.

1. The first is that the crisis of the welfare state, and in Western Europe the retreat from social democracy coinciding with the collapse of Communism in Central and Eastern Europe, have reduced both the enthusiasm for and the effectiveness of state planning. Of course, this is a highly complex (and still contested) phenomenon that cannot be adequately explored in this chapter. However, it may be misleading to assume that globalization has fundamentally changed the rules of the game, although Neave in chapter one is correct to emphasize that

the nation-state has ceased to be the sole framework within which higher education operates and that this is "no minor metamorphosis." The articulation between nation state and global context cannot be reduced to a zero-sum game. The state is still very much alive, as so many chapters in this book highlight. In certain important ways, the state is more active than ever. But it wields its influence in new ways, notably by espousing "strategies" as rhetorical devices as much as programmatic initiatives and by emphasizing its role as a regulator in the name of the public interest and as purchaser of public goods. The effect, however, has been to reduce the appetite for top-down planning and grand structural reforms (which both chimed in with the reaction against state planning in Central and Eastern Europe after 1990 but also may have weakened a necessary tool for transforming communist into post-communist systems).

Paradoxically, although the erosion of welfare-state politics and the rise of what has been called the market state have tended to problematize (if not undermine) the legitimacy of public higher education, they do not appear necessarily to have strengthened the legitimacy of private higher education. Indeed another conceptualization of the post-welfare state, the so-called regulatory state, may even have created new obstacles for private higher education:

(i) First, the state, now as a regulator rather than provider and/or funder of higher education, continues to control the legal environment within which private institutions must operate. Arguably, it exercises closer surveillance of what I prefer to describe as "post-public" higher education, that is private institutions and public institutions which have been granted greater operational autonomy (and often positively encouraged to behave in quasi-market ways), than of traditional state or public-sector higher education systems (which, in effect, were part of the state's own apparatus). Furthermore public institutions are perhaps in a better position to withstand, and respond to, this surveillance—because, from long experience, they understand the rules of the game better than private institutions. This is a particular problem in Central and Eastern Europe because the private higher education represents the "shock of the new," an apparently rapid and uncontrolled deviation from previous norms (although as Levy points out in chapter thirteen, the prominence of private higher education in the region is highlighted, even exaggerated, by its proximity to Western Europe, the region where private higher education is least well developed).

In contrast, in the United States where private higher education has the greatest legitimacy, stable and respected private institutions understand the rules of the game established by the regulatory state as well as public institutions;

(ii) Secondly (and more specifically), the regulatory state places greater emphasis on quality and standards that are the main object of regulation (on behalf of the "customers" of "post-public" higher education). Once again private institutions may be at a disadvantage—because of their more limited stake in research (often a key proxy for academic standards), their dependence on part-time teachers employed in public institutions and their concentration on particular subjects such as business and management, and computing, and information technology rather than teaching a more comprehensive range of disciplines. The preceding chapters demonstrate that these features are especially characteristic of private higher education in Central and Eastern Europe. For example, Giesecke in chapter four emphasizes the links between institutional effectiveness, institutional viability, and institutional legitimacy—links that apply to all higher education systems. But the comparatively weak institutionalization of many private colleges and universities, as demonstrated by the dependencies outlined above, may help to explain the problems they have encountered in terms of acquiring sufficient legitimacy. As Suspitsin points out in chapter seven, the (regulatory) state's supply-side regulation of academic quality is often at odds with the demand-driven imperatives of the market. In chapter three, Pachuashvili also emphasizes that regulation, while conferring legitimacy, may also inhibit growth and adaptability, qualities that private institutions must be able to demonstrate to survive and thrive.

Although it is possible that other forms of legitimacy, independent of state regulation, are emerging—for example, success in the market and endorsement by employers or students –this is only happening gradually in the region. Yet the rapid and at times chaotic expansion of private higher education following the collapse of the former communist regimes may even have made it more difficult for these alternative nonstate forms of legitimacy to become securely established. Several chapters—notably chapters six and nine on Romania by Reisz and Nicolescu respectively, and chapter eleven on the Ukraine by Stetar, Panych, and Tatusko—discuss these dilemmas in greater detail. In chapter ten on the public image of private universities and colleges in Bulgaria, Boyadjieva and Slantcheva directly address the emergence of nonstate forms of legitimacy. They argue that, although legal forms are important in determining legitimacy, so too are adherence to social norms and professional recognition. It is possible that these civil society sources of legitimacy may be as important as market sources in complementing, or competing, with state-determined sources—even for private institutions that are rooted in specifically market exchanges.

2. The second reason is that the growing emphasis on the knowledge society and globalization, as much as intellectual discourses as real-world trends (which tend to be less dramatic than these discourses), has presented differentiation in higher education systems in a new light. Because these discourses have introduced a new vocabulary—of instantaneity, volatility, ambiguity—attention has switched from the structuring of systems to the responsiveness, even the reflexivity, of institutions (Nowotny et al. 2001). Enter the market—but often in a state-sponsored, highly generic, even symbolic form rather than in terms of the more pragmatic practices familiar to private higher education institutions (which may help to explain the curious fact that these private institutions have not always benefited from this new emphasis on the market; they have often remained confined to the margins of systems). The focus in many countries was on mainstream, that is, public, higher education.

In the United Kingdom, for example, intermediary agencies beginning with the University Grants Committee and ending with the three national higher education funding councils (for England, Scotland, and Wales) switched from being buffer bodies designed to insulate universities from the excessive attentions of the state to become contracting bodies that communicated political (and socioeconomic) demands to higher education and introduced tighter accountability regimes. In the rest of (Western) Europe, significant reforms have taken place in the governance of universities, designed with the intention of freeing universities from the detailed tutelage of ministries—but not to increase institutional autonomy and academic freedom in their own right; rather to enable universities to behave in a more responsive and entrepreneurial manner. The Bologna process itself can plausibly be regarded as an exercise in marketization—internally to make European universities more flexible and adaptable; and externally to increase the global competitiveness of the 'European higher education area'. But, paradoxically perhaps, the development of private higher education is not regarded as relevant to this process of marketization. As Kweik warns in chapter five, the extension of the Bologna process to Central and Eastern Europe may actually inhibit the development of private higher education because of its dominant focus on the reform of public higher education systems.

It is in this context of this wider differentiation of institutional missions that the development of private higher education can be situated. In Western Europe, this development has been paltry—partly because of the prestige and resilience of public higher education systems, dominated by traditional universities with centuries-old stores of cultural capital; but also partly because this differentiation has often been state-sponsored if no longer state-planned. In Central and Eastern Europe, for reasons that will be examined in greater detail in the next section, the advance of private higher education has been more significant—partly because the prestige of public institutions had been compromised by their association with communist rule (while private institutions, previously forbidden, could start with a clean sheet); partly because public institutions sometimes associated market responsiveness with the subordination to political and social imperatives from which they had just escaped; partly because public higher education systems were inflexible and hidebound (and often chronically underfunded); and partly because the attitudes of post-communist regimes to the market veered wildly from the suspicious to the piratical (while in the West, the relationships between the state, civil society and the market were more orderly).

However, it may be a mistake to place too much emphasis on the strength or weakness of private higher education as an indication of the willingness or reluctance of national systems to embrace the market. Outside Central and Eastern Europe, this was a side-show—which raises the interesting question of whether higher education systems in that region are trendsetters, in the sense that private institutions will come to play a more central role, all differentiated systems; or whether the significance of private institutions in these systems is a passing phase attributable to the special circumstances surrounding the transition from communist to post-communist regimes. This is not a question that can be answered with the available evidence. But it may be that the stronger development of private institutions in Central and Eastern Europe is a response to these particular political circumstances (and so an internal phenomenon) rather than part of a wider response to the challenges posed by the emerging knowledge society and gathering pace of globalization that are driving differentiation at the system level and reflexivity at the institutional level in modern higher education systems (and which could be described as an external phenomenon).

This contrast is clearly relevant to any discussion of the legitimacy of these systems. While the development of private institutions in Central and Eastern Europe may pose particular problems in this respect, there are wider issues about the legitimacy of all higher education systems. Until the recent past, this legitimacy was grounded in a sense of "publicness"—whether because of the role played by universities in the development of national and cultural identities, or their role in providing quasi-democratic opportunities for higher learning, or their role in promoting socioeconomic well-being (and many private institutions, because of their reliance on philanthropy and their social and professional prestige shared in this sense of publicness). Today the context is different—and more contested. The development of differentiated systems raises awkward questions about access and social equity, especially if they also tend to be hierarchical systems. The dynamics of the knowledge society throw up equally difficult questions about intellectual property (and its relationship to the public good, to academic culture and to "open" science). Reforms in the governance of universities also raise difficult questions about institutional property rights. Debates about tuition fees, which inevitably involve discussions about the proportionality of individual and social benefits, lead to disturbing questions about the ultimate purposes of higher education. The need to reground the legitimacy of these more open, mass, market-like higher education systems (and, perhaps, of Mode-2 knowledge production systems too) is urgent and obvious—and extends far beyond the legitimacy of private institutions within a particular historical and regional context.

The Development of Post-Communist Higher Education Systems

Higher education systems in Central and Eastern Europe during the communist period had a number of distinctive features, some of which have continued and some disappeared (Scott 2000).

1. The first, but not necessarily the most significant, was that they were subordinated to governments in which the Communist Party monopolized power. This feature was not necessarily the most important for two main reasons. First, the opening of the archives has demonstrated the stubborn persistence of pluralism within the context of one-party rule; in the Soviet Union, for example, in the 1960s and 1970s, there were vigorous and sophisticated debates about economic reform (which echoed not dissimilar debates within Western Europe welfare states). Secondly, although political dissidents and ideological opponents were repressed in (or excluded from) higher education, the organization of institutions and the structure of systems were not substantially modified by the communist rule; remarkably little changed with traditional universities continuing to be firmly in the Humboldtian mold (with a Marxist-Leninist veneer) and higher professional schools that owed much to the Napoleonic/*grandes écoles* and the German technical university tradition; even the organization of

research, through institutes of Academies of Sciences, had its origins in the nineteenth century, firmly in the pre-communist era.

2. This organizational and institutional conservatism was the second distinctive feature of communist higher education systems. Long after Western European systems, often with similar origins, had been transformed by radical processes of democratization and massification, they were able to preserve more traditional patterns of organization. For example, as the binary distinction between classical universities and other less noble and more utilitarian higher education institutions came under increasing pressure in the West, it continued relatively unchallenged in Central and Eastern Europe—and not simply because these other institutions enjoyed relatively greater prestige or because (some would argue) the universities themselves had become more instrumental, deintellectualized institutions as a result of the imposition of communist ideology.
3. The third feature was that student growth had been less rapid in Central and Eastern Europe than in the West, a point many of the chapter authors find quite pertinent to the sharp private growth in the immediate post-communist years. More accurately, although growth rates from 1945 until the mid-1960s had been similar, later they accelerated in the West and stagnated in communist Europe (with the intriguing exception of the Soviet Union itself where they remained relatively high). As a result higher education had not really experienced the full force of massification before the collapse of Communism. There are a number of explanations for this divergence of growth rates. One is that student growth was regarded as politically problematical by the existing regimes; certainly events in Paris and elsewhere in the 1960s would have intensified their concerns that mass higher education might undermine ideological discipline. But a more plausible explanation is probably to be found in the divergence of economic growth rates between the East and the West from the 1970s onwards. The slowdown in student enrollment came at the same time as the stagnation of communist economies as they struggled unsuccessfully to come to grips with the challenges of an IT-fuelled post-industrial world.
4. The fourth distinctive feature, linked to this failure (and less highlighted in the preceding chapters), was the preservation of a balance between academic disciplines that had come by the end of the 1980s to be anachronistic. The emphasis on theoretical science may have reflected the materialism and utopianism of communist thought just as the emphasis on engineering reflected the priorities on the heavy industries in centrally planned economies. But preservation of the

> humanities as scholarly, even elitist, disciplines (rather than the mass disciplines they became in the West) and the underdevelopment of the social sciences may also have reflected political concerns about ideological subversion. But such motivations can easily be exaggerated. Preoccupation with the health of science and engineering in universities (and a desire to discourage the expansion of the humanities and, more critical, social sciences) were also commonplace policy motifs in the West.

However the collapse of communism left Central and Eastern European higher education systems with a number of deficits, whatever their causes. First, they were, not surprisingly, intellectually timid (even opponents of the regime displayed a degree of intellectual conservatism that had passed out of fashion in the West); the brilliance of Soviet scientific research in particular disguised more mundane deficiencies. Secondly, they were organizationally rigid, at both systemic and institutional levels. Despite (or perhaps because of) their strict subordination to the one-party state, they had failed to develop the flexibility fuelled by multiple transactions with multiple stakeholders which, however unwelcome to some, had become routine in the West; they were, in more than one sense, more closed systems. Third, their capacity was not sufficient to meet student demand, which for obvious reasons expanded rapidly after the collapse of communism; they had missed out on mass higher education, a defining rite of passage for advanced higher education systems elsewhere (Scott 1995). Finally, their subject balance was no longer appropriate in terms of both present and future student demand but also the needs of society and the economy with substantial overprovision in natural sciences (and, of course, Russian) and equally substantial underprovision in the social sciences (where management led a covert life as a branch of Marxist economics).

In the decade and a half since the collapse of communism there has been a significant development of private higher education—but only in part, to remedy some of these deficiencies. However, this development has not been consistent across the region. Paradoxically the advance of private higher education has been most limited in the most Western states—notably the Czech Republic and Slovenia. It has been most rapid in Poland where the number of private institutions now exceeds that of public institutions, although not the total number of students because many of the private institutions enroll small numbers. There are a number of possible reasons for this uneven development. Some relate to political circumstances external to higher education.

One possible factor may have been the political composition of post-communist governments, many of which have continued to be heavily influenced by sociodemocratic/post-communist forces. However, in practice there is little correlation between the position of governments on a right-left spectrum and the willingness to embrace private higher education. For much of the period, Hungary has been ruled by centre-left governments; yet this has not discouraged the development of private institutions (Poland, although now ruled by a centre-right government, was in a similar position for the majority of the post-communist years). Conversely, the Czech Republic, with more mixed governments, has proved to be an uncongenial environment for the development of private higher education. A second, more geopolitical factor may have been the orientation of post-communist higher education systems. In the Czech Republic and Slovenia, the orientation was firmly to Western Europe systems composed overwhelmingly of public institutions; in Poland and Hungary, the orientation has perhaps been more to American higher education where private institutions are more prominent. This orientation may be explained to some extent to the location of anticommunist diasporas before 1990.

However, there are other reasons more directly linked to the development of post-communist higher education systems. The most powerful being the pressure to expand these systems—but against a background of highly constrained public funding (and, in some cases, the reluctance of public institutions to expand because of their doubts about the—relatively unfamiliar—processes of massification). Both Pachuashvili in chapter three and Reisz in chapter six highlight the implications of this (quantitative and qualitative) underdevelopment of higher education systems during the communist era. Private institutions have been established to meet demands that could, or would, not be satisfied by public systems. This may help to explain the unevenness of development. In those countries with higher participation rates (for example, Russia), there has been less unmet demand—and in countries with higher GDPs and faster growth rates (for example, Slovenia) it has been more feasible to meet this demand by expanding public higher education. Conversely, in countries with low participation, low GDPs and low growth rates (for example, Romania), the development of private institutions has played a key role in meeting unmet demand.

A second, linked, reason has been the pressure to build up capacity in subject areas that were either actively discouraged during the communist period (for example, business and management) or had been poorly developed because of structural factors (for example, computing). Although these two subject areas are well (over?) represented in private higher education worldwide, there have been special reasons for their dominance of private higher education in Central and Eastern Europe. However, a third reason seems to have been less important—the need to establish higher education institutions untainted by collaboration with the previous communist regimes (which all public institutions, to some degree, had been). This experience contrasts with that of Latin America where some important private universities have been established partly to avoid the excesses of massification and early Catholic ones also to resist the secular, liberal, leftist values which, it has been argued, had infected public institutions.

The prospects for private institutions in post-communist higher education systems in Central and Eastern Europe are difficult to assess. Clearly a significant private sector will continue, if only because there were virtually no private institutions in the region before 1990 (only a few highly policed religious institutions). But it is unclear at what level the contribution of the private sector will settle—will it stabilize or will it continue to expand. Higher education systems in the region are going through a complex process of normalization. It is a complex process because it has several strands. One strand is the impact of the Bologna process that has been framed in the context of higher education in Western Europe (hardly surprisingly; the four signatories of the forerunner Sorbonne declaration were Britain, France, Germany, and Italy). In terms of its overall ethos, Bologna is a public (even state-dominated) process; and in its detailed elements, particularly its heavy emphasis on quality culture, it may curb the more entrepreneurial, even piratical, practices of some private institutions in Central and Eastern Europe.

But another strand is the continuing differentiation of higher education systems and growing diversity, even reflexivity, of institutions as the result of the emergence of a global knowledge society. Here practices developed in Central and Eastern Europe since 1990 may provide models for other systems—for example, the mixing of state-funded and fee-paying students in the same institutions, which, as a result, will take on a hybrid public-private character (although this is also a feature of the Australian higher education system where universities are allowed to admit full-fee-paying students over and above so-called HECS (Higher Education Contribution Scheme) students who receive state loans to pay their tuition fees). But, once again, it is possible that post-communist systems in Central and Eastern Europe may be influenced by Western European policy norms where the debate about tuition fees is different, in effect a gradual, reluctant, and often contested advance to higher fees for all (with compensatory support systems). A third strand of normalization, of course, is the extent to which all the new forms of higher education that are emerging (and which I prefer to label post-public rather than private) can successfully generate new forms of legitimacy. Here the experience of building post-communist higher education systems in Central and Eastern Europe—and, in particular, the prominent but contested role played by private institutions—offers a fascinating (but perhaps exceptional) case study.

Bibliography

Cloete, Nico. (ed.). 2002. Transformation *in Higher Education: Global Pressures and Local Realities in South Africa. Pretoria.* Sunnyside: Centre for Higher Education Transformation.

Levy, Daniel. 1996. *Higher Education and the State in Latin America: private challenges to public dominance.* Chicago: University of Chicago Press.

Lewin, Moshe. 2005. [edited by Gregory Elliott] *The Soviet Century.* London: Verso.

Nowotny, Helga, Peter Scott, and Michael Gibbons. 2001. *Rethinking Science: Knowledge and the Public in an Age of Uncertainty.* Cambridge: Polity Press.

Scott, Peter. 1995. *The Meanings of Mass Higher Education.* Buckingham: Open University Press.

——— 2000. Ten Years On: Higher Education in Central and Eastern Europe. Bucharest: UNESCO-CEPES.

Index

SNEJANA SLANTCHEVA is Head of the Center for Research on Higher Education in Central and Eastern Europe, Blagoevgrad, Bulgaria, and Head of PROPHE's Regional Center for Central and Eastern Europe. Her research and publications focus on higher education policy issues in the countries of Central and Eastern Europe. She has edited a multi-authored volume on private higher education in Central and Eastern Europe and has published articles and book chapters on higher education policy, the quality of teaching and learning, the academic profession, private higher education, and the Bologna process.

DANIEL C. LEVY is Distinguished Professor in the Department of Latin American, Caribbean and U.S. Latino Studies at the University of Albany, SUNY. He holds appointments in multiple units, including in educational policy. Levy directs the Program for Research on Private Higher Education (PROPHE), a global scholarly network funded by the Ford Foundation. His nine books and over one hundred articles concentrate on higher education policy, related non-profit sectors, or Latin American politics. Levy has lectured at top-ranked U.S. universities and has consulted widely on education policy in Latin America.

GPSR Compliance
The European Union's (EU) General Product Safety Regulation (GPSR) is a set of rules that requires consumer products to be safe and our obligations to ensure this.

If you have any concerns about our products, you can contact us on

ProductSafety@springernature.com

In case Publisher is established outside the EU, the EU authorized representative is:

Springer Nature Customer Service Center GmbH
Europaplatz 3
69115 Heidelberg, Germany

www.ingramcontent.com/pod-product-compliance
Lightning Source LLC
LaVergne TN
LVHW100253210926
848480LV00003B/910
9781349534821